Prioritizing Agricultural Research for Development

Experiences and Lessons

PRIORITIZING AGRICULTURAL RESEARCH FOR DEVELOPMENT

Experiences and Lessons

Edited by

David A. Raitzer

Center for International Forestry Research

and

George W. Norton

Virginia Polytechnic Institute and State University

CGIAR

ict-km
Communicate, Create, Collaborate
A PROGRAM OF THE CGIAR

CABI is a trading name of CAB International

CABI Head Office
Nosworthy Way
Wallingford
Oxfordshire OX10 8DE
UK

Tel: +44 (0)1491 832111
Fax: +44 (0)1491 833508
E-mail: cabi@cabi.org
Website: www.cabi.org

CABI North American Office
875 Massachusetts Avenue
7th Floor
Cambridge, MA 02139
USA

Tel: +1 617 395 4056
Fax: +1 617 354 6875
E-mail: cabi-nao@cabi.org

A catalogue record for this book is available from the British Library, London, UK.

Library of Congress Cataloging-in-Publication Data

Prioritizing agricultural research for development : experiences and lessons / edited by David A. Raitzer and George W. Norton.
p. cm.
Includes bibliographical references and index.
ISBN 978-1-84593-566-5 (hardback : alk. paper)
ISBN 978-1-84593-568-9 (pbk. : alk. paper)

1. Agriculture--Research--International cooperation. I. Raitzer, David A. II. Norton, George W. III. C.A.B. International.

S540.I56P75 2009
630.72--dc22 2009000112

ISBN-13: 978 1 84593 566 5 (hardback)
ISBN-13: 978 1 84593 568 9 (paperback – not for sale)

Typeset by SPi, Pondicherry, India.
Printed and bound in the UK by the MPG Books Group.

The paper used for the text pages in this book is FSC certified. The FSC (Forest Stewardship Council) is an international network to promote responsible management of the world's forests.

Contents

Contributors

Arega D. Alene, *IITA-Malawi, Chitedze Research Station, PO Box 30258, Lilongwe 3, Malawi. E-mail: a.alene@cgiar.org*

Sophie Alvarez, *CIAT, Km 17 Recta Cali – Palmira, Valle del Cauca, Colombia. E-mail: b.s.alvarez@cgiar.org*

Jock R. Anderson, *(formerly of IFPRI) 12894 Livia Drive, Catharpin, VA 20143, USA. E-mail: janderson@worldbank.org*

Cynthia Bantilan, *ICRISAT, Patancheru, Andhra Pradesh, 502 324, India. E-mail: c.bantilan@cgiar.org*

Aliou Diagne, *WARDA, Impact Assessment Economist, 01 B.P. 2031, Cotonou, Benin. E-mail: a.diagne@cgiar.org*

John Dixon, *CIMMYT, Km. 45 Carretera México-Veracruz, El Batan, Texcoco, Edo México, CP 56130, Mexico. E-mail: j.dixon@cgiar.org*

Boru Douthwaite, *CIAT, Km 17 Recta Cali – Palmira, Valle del Cauca, Colombia. E-mail: b.douthwaite@cgiar.org*

Mazen El-Solh, *ICARDA, PO Box 5466, Aleppo, Syrian Arab Republic. Email: m.elsolh@cgiar.org*

Keith O. Fuglie, *Economic Research Service, U.S. Department of Agriculture, 1800 M Street, N.W., Washington, DC, 20036, USA. E-mail: kfuglie@ers.usda.gov*

Peter Gardiner, *CGIAR Science Council Secretariat, C624, FAO, Viale delle Terme di Caracalla,00153, Rome, Italy. E-mail: peter.gardiner@fao.org*

J.D.H. Keatinge, *ICRISAT, Patancheru, Andhra Pradesh, 502324, India. E-mail: d.keatinge@cgiar.org*

Shellemiah Keya, *WARDA, 01 B.P. 2031, Cotonou, Benin. E-mail: s.keya@cgiar.org*

Patrick Kormawa, *WARDA, 01 B.P. 2031, Cotonou, Benin. E-mail: kormawa@yahoo.com*

Patti Kristjanson, *ILRI, PO Box 30709, Nairobi 00100, Kenya. E-mail: p.kristjanson@cgiar.org*

Roberto La Rovere, *CIMMYT, Km. 45 Carretera México-Veracruz, El Batan, Texcoco, Edo México, CP 56130, Mexico. E-mail: r.larovere@cgiar.org*

Ronald Mackay, *489 County Road Fifteen, Alfred, Ontario, K0B 1A0, Canada. E-mail: mackay.ronald@gmail.com*

Victor.M. Manyong, *IITA-Tanzania, C/o Mikocheni Agricultural Research Institute, Mikocheni B, Sam Nujoma Road, Plot 24B, PO Box 6226, Tanzania. E-mail: v.manyong@cgiar.org*

Ahmed Mazid, *ICARDA, PO Box 5466, Aleppo, Syrian Arab Republic. Email: a.mazid@cgiar.org*

Simon N'Cho, *WARDA, 01 B.P. 2031, Cotonou, Benin. E-mail: s.ncho@cgiar.org*

Jupiter Ndjeunga, *ICRISAT, BP: 12 404, Niamey, Niger. E-mail: n.jupiter@cgiar.org*

George W. Norton, *Virginia Tech, 205B Hutcheson Hall (0401), Blacksburg, VA 24061, USA. E-mail: gnorton@vt.edu*

David A. Raitzer, *CIFOR, Jl. CIFOR, Situgede, Sindang Barang, 16000 Bogor Barat, Indonesia. E-mail: d.raitzer@cgiar.org*

Thomas F. Randolph, *ILRI, PO Box 30709, Nairobi 00100, Kenya. E-mail: t.randolph@cgiar.org*

Robin Reid, *Center for Collaborative Conservation, Colorado State University, 1401 Campus Delivery, Fort Collins, CO 80523, USA. E-mail: Robin.Reid@ColoState.*edu

James G. Ryan, *18 Nungara Pl, Aranda, ACT 2614, Australia. E-mail: ryanjim@bigpond.net.au*

Diakalia Sanogo, *(formerly of IITA) IDRC – WARO, BP 11007 Peytavin, Dakar, Senegal. E-mail: dsanogo@idrc.org.sn*

Kamil Shideed, *ICARDA, PO Box 5466, Aleppo, Syrian Arab Republic. E-mail: k.shideed@cgiar.org*

Mahmoud Solh, *ICARDA, PO Box 5466, Aleppo,, Syrian Arab Republic. Email: m.solh@cgiar.org*

Graham Thiele, *International Potato Center, Apartado 1558, Lima 12, Peru. E-mail: g.thiele@cgiar.org*

Philip Thornton, (ILRI) *16 Mentone Terrace, Edinburgh EH9 2DF, UK. E-mail: p.thornton@cgiar.org*

Jamie Watts, *Bioversity International, Via dei Tre Denari, 472/A, Maccarese (Fiumicino), Rome 00057, Italy. E-mail: j.watts@cgiar.org*

Stanley Wood, *IFPRI, 2033 K Street, NW, Washington DC 20006, USA. E-mail: s.wood@cgiar.org*

Ousmane Youm, *(formerly of WARDA), Institute of Agriculture and Natural Resources, Department of Entomology, 202 Entomology Hall-East Campus, University of Nebraska, Lincoln, NE 68583-0816, USA. E-mail: O.youm2@unl.edu*

Foreword

Because of pervasive market failures, agricultural research, at least in many developing countries, is largely a public sector activity. By definition, this also means that there is no market to drive the allocation of funds to different types of research and across regions. Historically, these public sector funding decisions have been driven by a combination of scientific whim, political pressures and 'muddling through' (such as allocating budgets based on precedence). Even though the outcomes of research were widely recognized to be uncertain and long-term, in the 1970s economists began to propose methods for allocating public research funding in ways that would maximize expected social benefits. These methods were developed into quite sophisticated approaches that were eventually published in the landmark volume by Julian Alston, George Norton and Philip Pardey, *Science Under Scarcity*, Cornell University Press, Ithaca, 1995.

The International Agricultural Research Centers of the Consultative Group on International Agriculture Research (CGIAR), which at the time had considerable latitude to internally allocate their budgets, were well positioned to apply these methods. Through the 1980s, economists in several CGIAR Centers carried out research priority assessment exercises, many of which are referenced in this volume. However, the effects of these early efforts on actual Center resource allocations were not apparent, largely because they were seen as 'economists' exercises' by Center scientists and management.

Since the 1990s, a new generation of approaches to priority assessment has been used in the CGIAR Centers and partners. While the core economic methods have not changed significantly from the earlier efforts, the process has evolved considerably to include more participatory approaches that fully engaged scientists, research managers and external partners. At the same time, new methods such as geographic information systems and new databases such as poverty maps have added new tools to the core economic methods.

This volume provides a valuable and timely compilation of these experiences. In particular, it shows how a diversity of methods have been used to fit specific Center contexts (whether global and regional, or research-focused on genetic improvement or natural resources management), with more emphasis on simpler and more transparent methods that are readily understandable to a broader set of participants. Increasingly too, repeated priority assessment exercises in individual Centers and integration of *ex ante* and *ex post* impact assessment have enabled learning and feedback to be built into the process. It is heartening that this new generation of priority assessment is having more success in influencing research resource allocation decisions in the CGIAR Centers and partners, as illustrated in many of the chapters in this volume. Ironically, this has happened at a time when Centers have less flexibility to make these decisions due to a sharp increase in the share of their budgets that is from restricted 'project' funding.

The volume is also timely in taking stock of these rich experiences within the CGIAR and partners just when the CGIAR system is embarking on major reforms through a process of 'Change Management'. The final chapter provides a very useful synthesis of principles and current good practice to guide priority assessment for this 'new CGIAR'. Although much progress has been made, the volume also identifies many areas where methods and approaches need further development, including assessing non-market environmental costs and benefits, analysis of policy-oriented research, better targeting on the poor and the inclusion of indirect effects. Greatly increased uncertainty due to climate change, energy shocks and global market instability add additional challenges. Methods are continuing to evolve, especially with the growing availability of global databases and informational technologies to more sharply map agricultural systems and the poor. There is little doubt that with continued methodological progress and dedicated effort, better priority assessment will lead to greater impact on the ultimate beneficiaries of CGIAR and partner research, the poor of the developing world, thereby demonstrating the ultimate value of the experiences documented in this book.

Derek Byerlee
Co-Director of the 2008 World Development Report:
Agriculture for Development
Chair of the CGIAR Science Council,
Standing Panel on Impact Assessment

Acknowledgements

This book has benefited from the contributions of many individuals who generously donated their time to help inform and improve how research priorities are selected. The authors of each of the chapters tirelessly contributed chapters and revisions, often at short notice, without compensation, with the aim of helping to improve the impact potential of future agricultural research efforts. Without their perseverance, this book would have been impossible.

The Information Communication Technology and Knowledge Management Programme of the Consultative Group on International Agricultural Research generously contributed funding to the finalization and publication process through its Knowledge Sharing in Research Project. Nadia Manning Thomas, the project leader, provided valuable ideas for facilitation of the book's development, as well as complementary outreach methods.

Many support staff have helped to make this book a reality, both through assistance to individual chapters and through contributions to the overall volume. In particular, Lucya Yamin has helped to format, compile and harmonize the style of individual chapters, check references, compile the front matter for the book, as well as assist with other aspects.

We thank CABI for their assistance in copy-editing and publishing the work, and particularly appreciate Nigel Farrar's patience and cooperation as we finalized these contents.

Acronyms

AARINENA	Association of Agricultural Research Institutes in the Near East and North Africa
ACIAR	Australian Centre for International Agricultural Research
ACSAD	Arab Center for the Studies of Arid Zones and Dry Lands
AEZ	Agro-ecological Zone
AP	Arabian Peninsula
ARI	Africa Rice Initiative
ASARECA	Association for Strengthening Agricultural Research in Eastern and Central Africa
BCR	Benefit/Cost Ratio
CAADP	Comprehensive Africa Agriculture Development Programme
CAC	Central Asia and the Caucasus
CAPSA	Center for Alleviation of Poverty through Secondary Crops in Asia
CCER	Center Commissioned External Review
CCF	Consolidated Conceptual Framework
CD	Committee of Directors
CGIAR	Consultative Group on International Agricultural Research
CIFOR	Center for International Forestry Research
CIHEAM	Centre International de Hautes Études Agronomiques Méditerranéennes

CIIFAD	Cornell International Institute for Food, Agriculture and Development
CIMMYT	International Maize and Wheat Improvement Center
CIP	International Potato Center
COM	Council of Ministers
COMESA	Regional Economic Development Authority (COMESA)
CPs	Challenge Programmes
CPWF	Challenge Programme on Water and Food
CWANA	Central and West Asia and North Africa
DALYs	Disability-adjusted Life Years
DREAM	*D*ynamic *Re*search *E*v*A*luation for *M*anagement
ECA	Eastern and Central Africa
ECARRN	East and Central Africa Rice Research Network
EMM	Economy-wide, Multi-market Model
EPMR	External Programme and Management Review
FAO	UN Food and Agriculture Organization
FARA	Forum for Agricultural Research in Africa
GDP	Gross Domestic Product
GFAR	Global Forum on Agricultural Research
GIS	Geographic Information System
GTZ	Deutsche Gesellschaft für Technische Zusammenarbeit
HPI	Human Poverty Index
IARCs	International Agricultural Research Centres
ICARDA	International Center for Agricultural Research in the Dry Areas
ICBA	International Center for Biosaline Agriculture
ICLARM	International Center for Living Aquatic Resources Management (now WorldFish Center)
ICRISAT	International Crops Research Institute for the Semi-arid Tropics
IFAD	International Fund for Agricultural Development
IFPRI	International Food Policy Research Institute
IGP	Indo-Gangetic Plains
IITA	International Institute for Tropical Agriculture
ILAC	Institutional Learning and Change Initiative

ILRI	International Livestock Research Institute
IPG	International Public Good
IPGRI	International Plant Genetic Resources Institute (now Bioversity International)
IPM	Integrated Pest Management
IPs	Impact Pathways
IRR	Internal Rates of Return
IRRI	International Rice Research Institute
iSC	interim Science Council
ISNAR	International Service for National Agricultural Research
ITFFR	International Task Force on Forestry Research
IVC	Inland Valley Consortium
M&E	Monitoring and Evaluation
MTP	Medium-term Plan
N_2O	Nitrous Oxide
NA	North Africa
NARES	National Agricultural Research and Extension Systems
NARI	National Agricultural Research Institutes
NARS	National Agricultural Research Systems
NEC	National Experts Committee
NEPAD	The New Partnership for Africa's Development
NERICA	New Rice for Africa
NPPs	Networks, Programmes and Projects
NPV	Net Present Value
NRM	Research on Natural Resource Management
NVRS	Nile Valley and Red Sea
OSS	Sahara and Sahel Observatory
PAAP	Policy Analysis and Advocacy Programme
PAE	Priority Assessment Exercise
PIPA	Participatory Impact Pathways Analysis
PPP	Purchasing Power Parity
PRA	Participatory Rural Appraisal

PRE	Priority Ranking Exercise
R&D	Research and Development
RAT	Resource Allocation Tool
REIA	Research Evaluation and Impact Assessment
ROCARIZ	West Africa Rice Research Network
RRA	Rapid Rural Appraisal
RWC	Rice–Wheat Consortium
SACCAR	Southern Africa Center for Cooperation in Agricultural Research and Training
SADC	Southern African Development Community
SC	Science Council
SOs	Strategic Objectives
SP	Strategic Plan
SPs	System Priorities for Research
SROs	Subregional Agricultural Research Organizations
SSA	Sub-Saharan Africa
TAC	CGIAR Technical Advisory Committee
TFAP	Tropical Forestry Action Plan
TFPS	Task Force on Priority Setting
TOs	Tactical Objectives
TPS	True Potato Seed
UN	United Nations
UNDP	United Nations Development Programme
UNESCO	United Nations Educational Scientific and Cultural Organization
WA	West Asia
WARDA	West Africa Rice Development Association
WECARD/CORAF	West and Central African Council for Research and Development
WPW	Work Planning Week

1 Introduction to Prioritizing Agricultural Research for Development

David A. Raitzer and George W. Norton

Abstract
The rationale for systematic assessment of priorities for agricultural research for development is explained and introduced in this chapter. A brief overview is provided on the evolution of agricultural-research-for-development portfolios and institutional contexts, particularly the expansion of natural resources and policy-oriented research, the increasing number of goals that such research is intended to address and the recent shifts from core funding to competitive grants. The implications of these shifts for priority-setting methods are described and are related to the goals of the present volume. The introduction concludes by defining the intended audience for the book and introducing the content of the subsequent chapters.

Keywords: Priority setting, priority assessment, returns to research, international agricultural research.

Why Assess Research Priorities?

Sharp increases in food prices over the past year (2007/08) have drawn attention to the need to invest in agricultural[1] research. This need has been reinforced by recent recognition of the potential effects of climate change on agriculture. However, public sector budgets for international agricultural research are limited, so it is crucial that these scarce resources be invested effectively. Accordingly, it has been recognized that international agricultural research needs to be 'changed' to be more coherently focused on topics with potential to benefit the poor and/or the environment (CGIAR, 2008).

[1] For simplicity, 'agriculture' is used in this document as a comprehensive term for all natural resources management activities, including for forests and aquatic resources.

While agricultural scientists have an intuitive understanding of the issues and problems that they are researching, their research preferences are subject to a number of influences and limitations that often cause them to pursue areas that may not maximize impact potential. Like any other group, scientists are motivated by socially established norms and values, and they respond to incentives created by peer pressures. These incentives principally are oriented towards rewarding certain kinds of innovation and productivity, and often bias scientists towards upstream fundamental research, which is affiliated with 'prestige', or towards topics that are 'trendy'. While many scientists and managers promoted on the basis of scientific 'prestige' may understand how to select topics that maximize the uptake of their products among immediate users, most scientists do not have a clear understanding of indirect and long-term impact pathways, for which forecasting may require ambitious assumptions. Nor does such understanding necessarily translate into an ability to make decisions across the work of different scientists, which may require different sets of scientific skills.

Thus, in the absence of market forces, systematic empirical analysis (priority assessment) is needed to help guide limited resources to those research areas that have the greatest potential to produce benefits in line with research system goals such as productivity enhancement, poverty reduction and environmental improvement. Dedicated work is needed to help tease out implicit assumptions about the intended use of research products and to translate forecasted use into comparable metrics that represent contributions to different goals and objectives of a research organization. This work can help to build a portfolio that maximizes overall impact potential.

However, it can also do much more. Priority assessment can be an important 'learning' tool, as the assumptions elicited regarding the future use of research products can be tested against the results of past evaluations and experience, so as to enlighten and educate scientists. In addition, priority assessment can be an important way to engage partners and external stakeholders in a process of shared reflection over the validity of assumptions underpinning impact pathways. Priority assessment can also be a critical tool for monitoring and evaluation, as the process can be used to establish benchmarks for appraisal of progress and the validity of assumptions.

The Changing Context for Assessing Priorities

For many of the above reasons, international and national agricultural research agencies have regularly hired social scientists to provide information that will help guide research decision making. From early in the history of the Consultative Group on International Agricultural Research (CGIAR),[2] there was explicit attention to the development of economics and social science expertise, so as

[2] For example, CIMMYT (1970) and TAC (1973) explicitly refer to the roles of economists in helping to inform research choices, so as to maximize impact potential.

to help orientate research towards topics with maximum impact potential. Although a number of other research agencies also seek to use such information to help make transparent decisions about activities, such use is relatively unique for a public sector international organization.

Considerable effort has been invested by the CGIAR International Agricultural Research Center (IARCs) and the broader agricultural research community to develop systematic and transparent methods for identifying research topics with the greatest impact potential. These methods typically rely on *ex ante* indices of expected economic, poverty and environmental benefits and attempt to explore trade-offs among intended objectives. In turn, these estimates rely on articulation of 'impact pathways', including consideration of the probability of research success, likely patterns of adoption and indirect effects. At the same time, these efforts have been subject to strong – and growing – limitations.

Decisions over the allocation of research resources are increasingly fragmented, as research funding has shifted from core funds to competitive grants. For example, in the CGIAR, between 1995 and 2007, unrestricted funding fell from 63% of the funds received to 39%, and restricted project funding grew in inverse proportions (McAllister *et al.*, 2008). This increased importance of restricted funding means that the power to select and prioritize research activities largely shifted from a small number of IARCs to a much larger number of loosely coordinated donors.

Further fragmenting decision making is a parallel expansion of international agricultural research entities, even as the total funds available for research are stagnant. Between 1990 and 2007, CGIAR funds remained roughly level in real US dollars adjusted for inflation. However, the CGIAR went from a system with 13 IARCs focused on food production systems with two small support offices to 15 IARCS expanded to include forestry, fisheries and aquaculture, four Challenge Programs (with IARC-size budgets), 19 system-wide programmes and an expanded array of support offices. Meanwhile, the CGIAR and its Centers' mission evolved from a focus on enhanced productivity of the staple foods that the poor consume, to include environmental protection and a broader array of poverty-related goals.

At the same time, reporting and transaction costs have increased. For example, medium-term plans went from a five-year cycle to three-year cycle, to a plan that is 'rolling', with annual updates. With rising restricted funding, individual donors required greater reporting for specific grants, while the introduction of a complicated 'performance measurement system' in 2005 increased reporting burdens for core funding. The result has a context for priority setting that is evermore complex, with a greater array of objectives to be compared, but with fewer resources and opportunities for in-depth analysis of impact potential. These trends are by no means unique to the CGIAR, as many institutes within National Agricultural Research Systems (NARS) have been subject to similar trends. In this context, the CGIAR Center have still attempted to employ novel and systematic approaches to priority assessment, but the findings and methods of these efforts have often been subsumed by competing forms of information.

Meanwhile, the nature of the research portfolio of the CGIAR and its partners has fundamentally changed. While investment in disembodied information such as management recommendations, policy analyses and upstream understanding has grown, investment in embedded technologies such as improved varieties, biocontrol vectors and agricultural inputs has declined. By 2003, it is estimated that more than half of the IARCs' funding is now oriented towards these disembodied areas (SPIA, 2006).

Priority assessment methods, supported by economic surplus analysis, are relatively well developed for embedded technologies (see Alston *et al.*, 1998 for a comprehensive discussion), where logistic diffusion curves can forecast adoption with relatively few parameters, and adoption has predictable productivity effects. However, innovation is needed for the other areas of the research portfolio of the CGIAR and its partners, which have impact pathways that are more difficult to predict and benefits that are more difficult to quantify. Moreover, these methods will need to be appropriate to circumstances where scientists and managers may already feel 'planning fatigue' from existing reporting requirements.

Objectives of this Book

The intent of this book is to help bring the focus back to how decisions are made regarding the use of research resources. One role of this book is to illustrate that research institutes have the capacity to systematically assess the alternative uses of research resources, and that this capacity can be effective, particularly if appropriate incentives exist for its utilization. In so doing, a parallel goal is to help identify methods and modalities appropriate to tackle this new, more complex environment, with new areas of research, a broader array of objectives to be fulfilled and fewer resources available.

While there have been many prior priority-setting manuals (e.g. Alston *et al.*, 1998; Mills, 1998), no other source documents the breadth of real-world experience with priority assessment found in this book. The intention is to offer a unique menu of options and lessons for future research planning efforts by NARS institutes and their international partners. Unlike previous manuals, the focus here is not intended to be exclusively on methodology, but also on how the complexities and challenges involved in actual implementation have been broached, such as choosing an appropriate unit of analysis and eliciting parameter estimates for research use that are realistic from scientists. The intention here is not to present polished versions of perfect exercises. Rather, it is to show which aspects of methods worked well, which worked less well, and to identify which contributed and inhibited contribution to learning, coalition building and ultimately buying in of shifts in research agendas through case examples in different circumstances, under 'real-world' constraints.

In this book, two related, but different, terms are frequently used. One is 'priority assessment', which refers to systematic analysis of potential research alternatives. Another is 'priority setting', which refers to how actual decisions are made regarding the portfolio of research activities pursued by an institute, as well as relative resource allocation among them. Although every institute sets priorities, the reliance of 'priority setting' on 'priority assessment' has been variable. This book is intended to move beyond a focus on 'priority assessment' methods to reflect on how those methods and the processes by which they have been institutionalized have led to different influences on priority-setting practice.

Structure of the Book

This book presents experiences and innovations with priority assessment methods in CGIAR and its partners at various levels, and with respect to a diverse array of research areas. Each chapter presents and appraises one or more methods that have been used to articulate, explore and assess impact pathways and research priorities. Subsequently, each chapter appraises experiences with the methods described, so as to communicate and share the strengths and weaknesses encountered for each approach. These chapters are followed by a 'synthesis' chapter that draws together real-world methodological lessons from the case chapters. More specifically, the book chapters are organized as follows.

Part I: Background tools

Chapter 2 presents Participatory Impact Pathways Analysis (PIPA). This approach defines key assumptions about intended partners, users and potential adoption constraints for a future research project, all of which are essential parameters for appraising the impact potential of research alternatives. It combines the results-oriented perspective of a problem tree with network maps for informing the development of 'an outcomes logic model', and has been successfully applied to define research plans for a number of research projects. The method also shows promise as a potential means of generating information for input into quantitative approaches to priority assessment.

Part II: Institute-level approaches

Chapter 3 presents a rigorous priority assessment exercise of the International Potato Center (CIP), which is one of the most advanced attempts to date to assess the potential impact of agricultural research options on poverty and health. This assessment is also an excellent example of a learning-oriented approach, in that it draws on *ex post* impact assessments to modify *ex ante* parameters. The chapter also makes important observations about limitations to the influence of the exercise on the CIP agenda.

Chapter 4 presents the evolution of priority assessment at the International Institute for Tropical Agriculture (IITA). Since 2001, the Institute has moved from a consultative qualitative priority identification approach towards more systematic and transparent methods. The first step in the evolution was a participatory exercise to identify criteria for the ranking of alternatives. After consolidation, these criteria were applied to shortlist outputs within research themes. A more in-depth quantitative approach was recently developed to explore benefits accruing to different groups in Nigeria from alternative research investments.

Chapter 5 presents the history of priority assessment and setting at the Center for International Forestry Research (CIFOR). To address an agenda that is principally policy-oriented, with goals often related to conservation and environmental protection, the Center's approaches have historically been consultative and qualitative. Recently, the Center has taken steps towards more systematic priority assessment through a structured ranking dialogue and multicriteria scoring exercise, which shortlisted a much longer list of research alternatives. The chapter discusses some of the factors that require careful attention in designing an appropriate approach to priority assessment.

Chapter 6 presents the evolution of priority assessment and setting at the International Crop Research Institute for the Semi-Arid Tropics (ICRISAT). The Institute has a long history of applying multicriteria scoring approaches for evaluating research options, which have been informed at times by economic surplus analysis, and characterization of adoption domains. Although the Institute continues to update a structured database on impact indicators, in recent years it has moved to simpler, qualitatively scored indicators.

Chapter 7 presents priority assessment approaches employed by the International Livestock Research Institute (ILRI) which follow a slightly modified version of those applied by ICRISAT. The chapter then describes how the exercise influenced the Institute's agenda in the 10-year period since the priority assessment exercise was completed – often in ways that might not have been predicted at the onset of the exercise.

Chapter 8 gives an overview of the methods and processes used in assessing research priorities at the Africa Rice Center (WARDA) during the past decade. This approach employed criteria associated with the Center's objectives, which were weighted by stakeholders. Notional research alternatives were developed with much explicit attention to the probability of success, dissemination and adoption. These alternatives were subjected to qualitative scoring in a workshop format, and were used to inform resource allocation decisions.

Chapter 9 describes the framework for priority setting and targeting at the International Maize and Wheat Improvement Center (CIMMYT). In 2004, the Center developed a 'resource allocation tool' (RAT) to systematically provide data for informing allocation decisions. The Center has also experimented with the use of geographic information systems and participatory methods for targeting of specific technologies.

Part III: System and regional approaches

Chapter 10 discusses a regional consultative approach for obtaining a shared vision of research priorities in Central and West Asia and North Africa (CWANA) which was led by the International Center for Agricultural Research for the Dry Areas (ICARDA). Consultations involved brainstorming meetings at subregional levels, a questionnaire distributed to stakeholders in the region, scoring and a final regional meeting. The emphases identified then helped to inform a new strategic plan for ICARDA.

Chapter 11 describes a data-intensive subregional approach taken by the Association for Strengthening Agricultural Research in Eastern and Central Africa (ASARECA), an organization mandated to coordinate and support agricultural research in the subregion, with support from the International Food Policy Research Institute (IFPRI). Using geographic information systems, information on agricultural potential, access to markets and population density was combined to delineate agricultural development domains. To illustrate the potential effects of productivity-oriented research, economic models were then used to simulate the effects of a 1% per year increase in productivity for a number of crops. This was then provided to stakeholders to guide discussion of decisions about investment across the eight identified development domains.

Chapter 12 describes the approach taken to define System Priorities for the CGIAR by its Science Council. The process followed was based on discussion and induction, rather than quantitative evaluation, informed by an array of background 'think pieces' and precedence analyses. By identifying 20 priority areas, but not ascribing relative priority among them, the intention was to provide a framework within which further planning for research implementation and resource allocation could take place by coalitions of CGIAR Center.

Part IV: Synthesis and ways forward

Chapter 13 synthesizes the ideas and lessons from earlier chapters. It identifies and highlights a few basic priority-assessment principles that can help improve consistency in methods across centers and programmes, followed by brief suggestions for priority-assessment methods that incorporate impacts of natural resource management and policy-oriented research.

References

Alston, J.M., Norton, G.W. and Pardey, P.G. (1998) *Science Under Scarcity: Principles and Practice for Agricultural Research and Priority Setting*. CAB International, Wallingford, UK.

CGIAR (2008) *Change Is in the Air: A Revitalized CGIAR – A New Way Forward*. CGIAR Secretariat, Washington, DC.

CIMMYT (1970) *CIMMYT: A Proposed Five Year Plan: 1971–1975*. International Maize and Wheat Improvement Center (CIMMYT), Texcoco, Mexico.

McAllister, E., Bezanson, K., Chadha, G., Mugabe, J., Waage, J., Perkins, K. and Watson, K. (2008) *Review of the Network/Partnership of the CGIAR, the Science Council and Affiliated Research Centers*. The World Bank, Washington, DC.

Mills, B. (ed.) (1998) *Agricultural Research Priority Setting: Information Investments for the Improved Use of Research Resources*. ISNAR, The Hague, The Netherlands.

SPIA (2006) *Impact Assessment of Policy-Oriented Research in the CGIAR: A Scoping Study Report*. Science Council Secretariat of the CGIAR, Rome, Italy.

TAC (1973) *Socio-economic Aspects of International Agricultural Research*. Technical Advisory Committee Secretariat of the CGIAR, Rome, Italy.

2 Participatory Impact Pathways Analysis (PIPA) and Research Priority Assessment*

BORU DOUTHWAITE, SOPHIE ALVAREZ, J.D.H. KEATINGE, RONALD MACKAY, GRAHAM THIELE AND JAMIE WATTS

Abstract

Participatory impact pathways analysis (PIPA) is an evolving tool that offers project managers a deeper understanding of the results that projects might attain with specific partners so as to help set priorities and support funding proposals. In a participatory manner, two groups of information are generated for each project. First, a problem tree is developed to represent the pathways by which research outputs are linked with outcomes and impacts. Second, network maps identify the key players and the roles they must play during and after each project to ensure its success. These two views of a project's impact pathways (IPs) are integrated in an outcomes logic model that describes what strategies the project will use to bring about the necessary changes, or outcomes, in project stakeholders to achieve the project vision. PIPA complements existing project management tools, such as the logical framework, by describing project strategies to bring about change, whereas traditional project planning instruments focus more on the activities required to produce the research outputs. Only when research outputs are used do they contribute to change. Hence, together with traditional project planning, PIPA provides information to allow priority assessment on the basis of scrutiny of the plausibility and the size of the envisaged change. PIPA can help to design projects to achieve overall programmatic goals and can help select between competing strategies within a single project. In the latter case, by mapping out potential IPs with a range of stakeholders, all partners are informed about the potential options considered. This common understanding is informative even if the final decisions on what the project will actually do are made by the project staff/leader.

Keywords: Participatory impact pathways analysis, priority setting, network maps, logic model, problem tree.

* The authors acknowledge valuable comments on drafts of this chapter by Douglas Horton and David Raitzer.

Introduction

Participatory impact pathways analysis (PIPA) is being used and refined by the Challenge Programme on Water and Food (CPWF), several Consultative Group on International Agricultural Research (CGIAR) Center, National Agricultural Research and Extension Systems (NARES) and other research for development (R&D) organizations. It is a young approach, which continues to evolve (see http://impactpathways.pbwiki.com). This chapter briefly summarizes the key elements in PIPA, its role in priority assessment, its uses in mid-term planning at CGIAR Centers, the experiences in using PIPA and the issues that have arisen as it has been applied.

Key Elements of PIPA

Underlying any research for development project are theories – more often implicit than explicit – about how the intervention is meant to change the status quo (Weiss, 1995; Rossi *et al.*, 2003; Chen, 2005). PIPA attempts to capture the theories of change for a given project, which can be understood as 'the pathways by which research outputs are most likely to yield impact' (CIFOR, 1996). The analysis helps research managers decide whether to fund the project by answering the key question: 'Given the project design, are the desired impacts likely to occur?'

Using a range of participatory techniques, PIPA develops four main products for any given project:

- a problem tree;
- a vision of what the project will help achieve, usually set at 2 years after the project finishes;
- network maps showing the current institutional context and as it would need to be in the future in order to achieve the vision;
- an *outcomes logic model* describing the changes in stakeholder knowledge and practice necessary to achieve the vision, and the strategies the project will use to bring about these desired outcomes.

As an additional fifth element, after the workshop an *impact logic model* can be developed describing how, by achieving its vision, the project believes it will contribute to longer-term impacts (such as livelihood improvement and poverty reduction). The impact logic model is usually complemented by a narrative that describes and explains the underlying outcomes logic, impact logic and the assumptions and networks involved.

Fundamental to PIPA is the idea that projects will only achieve their vision if at least some of the knowledge and technologies they produce are actually used. Hence adoption – the putting to use by actors in the innovation system in which the project operates – is fundamental.

PIPA distinguishes between two types of adoption – scaling-out and scaling-up. In the context of farm-level technologies and recommendations, scaling-out is the *horizontal process* of farmer to farmer diffusion, adoption and adaptation, often motivated by engaging stakeholders at a similar level to the original project. This can entail recruiting more sites where additional farmers and communities will further innovate and improve upon the R&D technologies introduced. Scaling-up is a *vertical process* requiring the engagement of a wider range of stakeholders (e.g. policy makers, government agencies, non-governmental organizations (NGOs)) at a higher level than the original project worked. These higher-level stakeholders command authority or influence to help build a more enabling environment in which the scaling-out process can thrive. For example, a new variety may be developed by a single organization but its use will depend upon a range of interdependent actors, including farmers, who value the variety and what it can do. Such actors would include organizations which make it widely available, regulatory bodies which exercise control over its quality and accessibility, governments which promote a suitable policy environment for its production and marketing, processors who convert it into food, and marketers who make it available to consumers, who finally purchase it.

Part of PIPA's originality lies in the use of network maps to represent such interactions both at the beginning of the project and at the scaling-out and scaling-up stages, when the existing networks need to change to achieve the project's vision. The outcomes logic model describes project strategies to achieve these changes. Hence, PIPA provides information that allows projects to be assessed on whether conjectured future scaling-out and scaling-up networks are consistent with achieving the project vision and whether the project strategies are likely to help bring about the identified changes. Furthermore, projects that have been through a participatory process to identify key stakeholders, how they are currently linked, and how a project fits in, are expected *a priori* to have a better chance of contributing to change than a project that has not, as assumptions concerning the impact pathways (IPs) are more explicitly reviewed. The case for funding a project is further strengthened when it can also demonstrate that it will regularly revisit its initial assumptions and monitor and evaluate its progress in forming and strengthening the research and the scaling-out and scaling-up needed to achieve impact.

The Role of PIPA in Priority Assessment

Essentially, priority setting involves choosing among alternatives at the project, programme or institute level. Priority assessment exercises typically require a defined set of technology or policy interventions for analysis (i.e. the research alternatives), grouped into reasonably coherent and discrete research projects that target well-defined constraints or provide opportunities to achieve practical impacts (Fuglie, 2007). Priority assessment exercises are also often carried out on non-research interventions that generate public goods including investments in capacity building and maintaining service units such as germplasm collection and conservation, and information units. An approach is then selected to assess the potential of the interventions to overcome one or more productivity constraints or to contribute to environmental or social objectives given the specified resources and time. Assessing the likelihood of each project to achieve its intended impact and determining the potential for adoption and support are critical steps in priority assessment (Walker and Collion, 1997).

The products of PIPA – the problem tree, vision, network maps and logic models – provide information that can help with two kinds of priority setting.

The first is internal priority setting within a given project. This is the kind of priority setting that selects which activities, series of actions and partners – from all those available – are those most likely to contribute to the outcomes and impacts that the project seeks to bring about. The second kind of priority setting – the kind most associated with the term and that is addressed directly in this chapter – is where project-funders make allocation decisions based on the prima facie expected merits of competing projects.

The challenge for priority assessment among research for development projects is to make valid comparisons between the merits of highly complex projects[1] or research projects with protracted IPs. This requires handling *trade-offs* between different dimensions of impact, e.g. between productivity and efficiency objectives versus poverty, equity or gender concerns (Bantilan and Keatinge, 2007). It also requires estimating the probability of success (likelihood of impact being achieved) and the likelihood of potential spillover effects into similar agronomic, climatological and ecological zones. Because of the high degree of uncertainty in this process, many assumptions have to be made. The basis for making these assumptions should be in consultation with different groups of stakeholders, and this requires weighting: for example, scientists are probably best placed to assess whether outputs can actually be delivered but extension workers and final users may be better placed to estimate the probability of adoption.

PIPA provides a method for the construction of project IPs with the stakeholders involved, which improves the reliability of the information about trade-offs among types of impact, likelihood of adoption and probability of spillover. Hence, undertaking PIPA has proved useful in adjusting existing projects and programmes. However, the participatory nature of PIPA and the lack of comparable quantitative metrics mean it is difficult to use as a tool to choose between competing projects. Nevertheless, it can help to inform research choices. For example, the CPWF uses problem trees and lists of actors as a criterion for selecting proposals for new projects when it makes a call. Regular updating of projects' IPs and relations to the programmatic IPs allow regular reflection on project and programme priorities and changes in them.

IPs Origins and Uses

PIPA builds on a number of methods. These include logical framework analysis (Suchman, 1962; Wholey, 1977), chain of events (Bennett, 1979), concept mapping (Yampolskaya *et al.*, 2004), action-to-outcome mapping (Jones and Seville, 2003), impact chains (Mayne, 2001), impact flow diagrams (Guijt, 1998), path analysis (Weiss, 1972), development pathways (Pender *et al.*, 1999), outcome engineering (Kibel, 1999), outcome mapping (Earl *et al.*, 2001) and results chains (Mayne, 2001, 2008b). PIPA shares many of their characteristics and draws on developments in the field of evaluation as well as on concepts from organizational learning and social network theory. Part of the novelty of the approach lies

[1] Succinct and valuable discussions of the differences between simple and complex settings and the challenges presented by the latter to project planning, implementation and evaluation can be found in Rogers (2008) and Mayne (forthcoming).

in its participatory nature coupled with the integration of *both* a causal-chain (i.e. problem tree) *and* an actor-orientated description of a project's IPs.

Within the R&D community, papers by Kuby (1999, 2000), Douthwaite *et al.* (2003 and 2007) and Springer-Heinze *et al.* (2003) have formulated, adopted and adapted the concepts that inform the use of IPs in the CGIAR system. Some donors also require analyses for funding proposals that share some characteristics with PIPA[2] and may even be prepared to invest more in a project pre-proposal phase[3], which includes workshops that bring together many partners to develop IPs, articulate roles and responsibilities and develop logframes.

The CPWF has been developing PIPA since October 2005 for use with its 50 or so projects. Some of the tools have also been used to help construct its medium-term plan (MTP). The International Potato Center (CIP) is using PIPA in a large British Government-funded project called Andean Change and is adapting it for reconstructing IPs of projects being implemented or already finished. The CGIAR Institutional Learning and Change Initiative (ILAC) used PIPA during the first meeting of its Learning Laboratory as a basis for developing a shared understanding and framework for action among the participating programmes. It is expected that the results from PIPA will serve as the basis for allocating funding through ILAC's small grants and for priority setting for technical assistance, training and other activities.

Use of IP analysis in MTPs

Recently, the CGIAR's Science Council (SC) has had evolving requirements for different IP-type analyses in CGIAR Center rolling MTPs three-year duration. These plans constitute a core planning and priority-setting document for all CGIAR Centers. Initially the SC required a logical framework (a type of logic model) to be prepared for each project. More recently, in 2006, project narratives were required that addressed elements of PIPA such as problem analysis, beneficiaries, partners and risks and assumptions, *in addition to* logframes. The 2007 MTP required an actual IP description, a project narrative and a logframe. The requirements of the SC for IPs have been a key force driving CGIAR Centers to experiment with the use of IPs in general and PIPA in particular.

Although the SC's version of IPs does not explicitly include the obligatory use of network maps, it increasingly recognizes the importance of the broad repertoire of roles that partners can and do play in project success. The different roles that CGIAR Centers as partners can play have been acknowledged and refined to include researcher, facilitator, enabler, catalyst and advocate. These roles are not limited to CGIAR Centers; indeed they are roles that are required in the IP of all development projects and can be played by a range of different actors.

In its instructions to CGIAR Centers for preparing their MTP, the CGIAR SC provides guidance on how to prepare an IP (CGIAR, 2007). In summary, the SC requires the description of IPs for each project 'from research outputs (reflecting problem identification) through outcomes to the ultimate impacts for

[2] For example, see the instructions in the Open Requests For proposals on the Bill and Melinda Gates Foundation Web site at http://www.gatesfoundation.org/default.htm

[3] For example, see the conditions set for the Project Preparation Grants on the Global Environmental Facility Web site at http://www.gefweb.org/

achieving CGIAR's goals' (CGIAR, 2007, p. 3). A project's IPs should also identify specific factors (e.g. policy and institutional constraints) that might inhibit achieving the outcomes and so limit the range or intensity of impacts. Each pathway should also identify target ecoregions, direct users of outputs and the end beneficiaries. It should describe the Center's role and also identify partnership arrangements and partner roles that are necessary to move from outputs to outcomes and impact. Finally, it should determine the capacity-strengthening needs of partners with capacity constraints.

In addition, the MTP includes a project logical framework that is given in table form as a simplified presentation of the IP. Thus, the reader of the MTP should see a direct correlation between the IP narrative and the logical framework presentation of outputs, users, outcomes and impacts. When Centers play a catalytic, facilitating, enabling or advocacy role, the IPs explain how these roles complement the Center's research role and contribute to the production of international public goods.

As the IP moves from outputs to outcomes to impact, research institutes have less control over the achievement of each step. Thus, while CGIAR Centers are expected to identify output targets and have some role in assuring outcomes, they are not expected to be able to predict the precise time frame or the extent of impact. However, CGIAR Centers are expected to provide evidence of impact and thus impact assessment–related activities should be included in IPs.

Experiences with IP Analysis

This section summarizes the experiences of Bioversity, CIP, International Crop Research Institute for the Semi-Arid Tropics (ICRISAT), CPWF and ILAC in their use of PIPA tools for a range of uses that include the development of MTPs and the use of PIPA to predict *ex ante impact*. To encourage this kind of reflection, we have followed an 'after action review' format (http://en.wikipedia.org/wiki/After_Action_Review). This format has allowed PIPA users to examine what worked well, what did not work so well and to identify improvements. An example of how PIPA has influenced research at the WorldFish Center appears in Appendix 2.1.

CGIAR Centers' positive experiences with IPs

CGIAR Centers and related research for development organizations are experimenting with the use of IPs. No one organization has yet developed a fully integrated approach, although PIPA is striving to provide one. Table 2.1 offers some informative lessons regarding the employment of IPs, based on discussion among research evaluators and planners from Bioversity, CIP, ICRISAT, CPWF and ILAC.

Some CGIAR Centers have tried to promote the use of information related to IPs for ongoing monitoring and reflection by transforming the information generated by the IP analysis process into data sets and managing it as such. Once the IP data are stored in electronic databases, they can be analysed to respond to emerging questions about the work (to reassess original project

Table 2.1. Insights gained from Bioversity International's, CIAT's, CIP's, ICRISAT's and CPWF's positive experiences with IPs, based on author reflections.

Positive experiences	Insights
Common understanding, sense of purpose, ownership and commitment to collective action are established among participating projects and programmes	The collaborative and participatory character of PIPA promotes project ownership. PIPA workshops held for CPWF projects in eight river basins proved one of the most successful mechanisms for forging programmatic integration. PIPA helped ILAC forge shared understanding and commitment to collective action in the projects and programmes participating in its Learning Laboratory.
IPs help provide focus, clarity and cohesion to individual research activities	The CGIAR SC in its review of the CPWF 2007–2009 MTP wrote: 'The CPWF has introduced the use of "objective trees" at the MTP project and CP level, a useful and innovative complement to the MTP logframe. In addition to providing a useful overview, the process of preparing these flow charts has clearly helped the CP provide the necessary focus, clarity and cohesion that now exists in the research plans at all levels'. CGIAR Centers have found that constructing IPs in the context of the MTP is helping them link activities to outputs, and outputs to outcomes.
PIPA has credibility with many stakeholders and partners	Researchers from both the 'hard' and the 'soft' sciences find PIPA credible and intuitively satisfying. So do other partners with diverse interests and backgrounds such as farmers and local politicians.
Participants in PIPA recognize network mapping as a powerful tool, that can handle issues of gender, institutions and power	Creating network maps focuses participants' attention on multiple aspects of partnerships. Network mapping is sensitive to – and so participants become aware of: (i) The very different but complementary roles that different partners play and the essential activities that they must undertake (ii) The importance and need for partners to reflect on the precise relationships that must hold between different partners (iii) The necessity of influencing these relationships – and how this is best accomplished – in cases where they are not already optimal (iv) Who holds power and influence in the networks, and who is being excluded.
Dedicated time is allotted to clarifying assumptions about impact	The requirement to develop IPs for the MTP or for a project is a means of helping to institutionalize the use of IPs, since institutes and projects must dedicate time for the exercise which otherwise might not be found. Assumptions which would otherwise remain implicit are made explicit, assessed and modified if necessary.

Continued

Table 2.1. Continued

Positive experiences	Insights
Impact orientation is promoted	The discipline of developing and thinking about IPs engenders and promotes an outcome and impact orientation among participants. Constructing MTP IPs has enabled CPWF projects to identify impact-oriented issues that cross borders of 'hard' and 'soft' science, since a thorough analysis of impact often requires projects to address policy, capacity and social and other issues. It has encouraged CPWF projects to think broadly and so consider impacts on technology, income, policy and institutions.
Facilitates discrimination between the efficacy of alternative courses of action within a project	During the PIPA process, alternative theories implying different IPs involving alternative courses of actions are revealed. Participants are therefore required to decide between these competing courses of action based on their perceived contribution to the impacts sought. In effect, this is a form of *internal priority setting* in which one course of action is preferred over another on the basis of its promise to deliver or contribute to impact.
IPs can help reveal multiple roles that may be necessary for impact	SC guidelines for the development of IPs recognize that, in order to influence the pathways, any given Center can play multiple roles in any given project, e.g. primary or secondary research provider, catalyser, facilitator, enabler and/or advocate to influence the pathways.

assumptions, to monitor progress of outputs, to query and to assess by different combinations of factors, etc.). This should provide a stronger basis for redirecting programmes to respond to new insights that emerge from such analysis and reflection. A preliminary example is the Bioversity and CGIAR effort to develop an online database for MTPs which store IPs as narrative text fields and as elements in a logframe matrix. However, it does not include the full details of PIPA, such as problem trees, network maps and assumptions. The information generated by PIPA for the ILAC cases has been made available to all participants as Excel spreadsheets and network maps. It is currently centrally stored and shared as e-mail files, and is accessed via an online portal that allows all team members equal access to all data. There is, however, an 'organizational culture' issue (Mayne, 2008a) since many CGIAR Center managers are biological or physical scientists who may not ascribe high value to such institutional information.

Difficulties Reported in the Use of IPs Analysis

All organizations contributing to this chapter report that they encountered some difficulties in their initial use of IPs. Difficulties, and potential solutions, tend to fall into two categories: methodological issues and implementation (Table 2.2).

Table 2.2. Issues identified while using PIPA methods in Bioversity International, CIAT, CIP, ICRISAT and CPWF via author reflections.

Opportunities/difficulties	Discussion and proposed solutions
Methodological issues	
PIPA elicitation tools are not yet fully developed	CGIAR Centers and CPs use different approaches to construct IPs for projects and programmes. PIPA is one such approach and the one we are recommending as part of priority setting. An emerging community of practice (see Table 2.1 above) is seeking to speed up the evolution of IP methodology and to develop guidelines for the use of IPs under different circumstances.
Constructing detailed IPs is time-consuming	The time devoted to the development of the IPs in the CGIAR context is often insufficient. Increasing the allocation of time would result in better internal vetting of projects and could increase the likelihood of engaging external partners in the development of the MTP.
Conflicting perspectives on IPs	Different perceptions exist regarding what an IP is, how an IP is best constructed and the precise purpose of an IP. Participatory construction of IPs in a workshop setting (e.g. PIPA) can help satisfactorily address this challenge.
Issues related to implementation	
Difficulty developing IPs for projects working in multiple locations and/or with many IPs	In the CGIAR context, MTP projects represent several million US dollars of annual investment, and they are actually programmes made up of a number of grants or projects. PIPA can be used at any level, but is easiest to use for individual projects in a single geographic locality. More upstream research activities with more potential and protracted IPs are more difficult to predict. The construction of programme IPs sometimes requires research scientists to be more explicit about assumptions than they are comfortable with.
Unequal levels of mastery/understanding	Staff members' levels of understanding and mastery of the process of producing IPs vary from rudimentary to skilled. Bioversity, in early 2009, is planning to carry out training on impact assessment that includes the construction of IPs, and has also conducted seminars on the topic for staff.
Linking IPs to planning, monitoring and evaluation and impact assessment	IPs provide testable hypotheses about what a project will change, and how. Hence, they provide a framework for monitoring and evaluation and *ex post* impact assessment. This potential is yet to be fully realized and is being actively pursued by the CPWF and ILAC.
Iterative use of PIPA	In the case of complex projects and programmes being carried out with multiple partners, it is likely that objectives evolve over time. It is important that PIPA be used in these cases in an iterative fashion (periodically over time), so that changing priorities, assumptions, network partners and other issues can be identified and managed.

As experience with PIPA reveals its strengths and weaknesses, and as these weaknesses are addressed, the component parts of the process are modified. This modification means that users must maintain a familiarity with how PIPA is being refined and why these refinements are necessary. The PIPA web site (http://impactpathways.pbwiki.com) has been created to serve this end, thus allowing users not only to keep abreast of modifications but also to share their experiences with the approach.

Realizing the Potential of IPs in Priority Assessment

A priori, we might expect more traditional quantitative priority assessment methods to require less consultative input when discrete projects are undertaken in relatively familiar contexts, where the networks of partners required for success are already defined, and there is a track record of impact (Ekboir, 2003). PIPA may add more value when these conditions are not met; for example, in newly established projects where partners still need to be selected or where IPs are not yet clearly conceived. Campbell *et al*. (2006, p. 53) state that priority assessment tools 'must help the innovation and associated decision making processes by assisting innovators to better see patterns and factors of success, trade-offs, and risks'. PIPA's participatory approach to mapping and analysing the theories underlying project activities and the activities of partner groups helps bring these trade-offs to the priority assessment process.

Based on the experience summarized in Tables 2.1 and 2.2, we have identified four key issues that must be addressed if PIPA and the use of IPs in general are to reach their potential to contribute to priority assessment and indeed to the broader field of R&D planning.

Issue 1: the appropriate unit of analysis for the construction of IPs

As discussed above, experience to date suggests that IPs are being constructed for different units of analysis – (i) the grant project[4] level; and (ii) the programme level – with varying levels of success. For the purposes of this chapter, we understand a project to be a single grant with fixed beginning and end dates, whereas a programme is made up of two or more projects, to meet a broader need over a longer time frame. At the individual project level, IPs are being used to identify how the project intervention is likely to influence outcomes and contribute to desired impacts. In addition to making project theory explicit, this use is also in effect an internal priority assessment exercise. A programme can use PIPA results to prioritize across projects by assessing trade-offs and by comparing probabilities of project success. Monitoring based on how projects' IPs unfold and evolve is a form of action research that adds to our understanding of how change happens (Mayne, 2001, 2008a). Both cross-project comparisons and IP evaluation contribute to external priority assessment.

[4] Note that 'project' here is defined differently than in the CGIAR MTPs discussed elsewhere.

Different stakeholders typically start out with somewhat different assumptions about how they expect to move from project activities to the impacts they seek. The participatory process involved in making these different assumptions explicit provides new information which often exposes the inadequacy of the overly simplistic. The process allows participants to converge on a more robust and better-informed theory. During this process they are faced with the task of determining which set of alternative activities within the same project is more likely to contribute effectively to desired impacts. Experience suggests that there are no major problems associated with conducting PIPA at the individual grant project level.

However, the use of PIPA at the level of programmes made up of multiple projects presents a greater challenge. Drawing network maps for programmes that work in many sites with many stakeholders or through many potential IPs is challenging. Indeed, there is a danger that, in extreme cases, it may prove frustrating and dissuade participants from further use. Furthermore, a programme is comprised of projects which may or may not harmonize with one another. When projects are not well integrated, it is not possible to develop coherent programme IPs without first clarifying what it is the programme as a whole is trying to achieve. IPs describe the underlying programme theory. If the programme theory is not coherent, then the IPs of its constituent projects may be irreconcilable.

Ideally, the programme-level IPs should be developed first to provide an overall goal for its projects. A PIPA process should be used as a way to involve key policy-level stakeholders and to ensure the programme is embedded and in concert with existing efforts to achieve similar types of changes. Project IPs can then be constructed in a similar way, followed by further detailing and adjusting of the programme-level IPs. The IPs are thus nested – like Russian dolls – with the programme IPs both forming and being formed by the project IPs. A similar concept of nesting has been used with traditional logical frameworks (Farrington *et al.*, 1997).

When a project is not contributing to the programmatic IPs, the contribution of that project to the programme as a whole is suspect and so should be examined carefully to determine whether or not it should or can be part of the programme portfolio. This examination is another form of priority assessment at the programme level.

Issue 2: the need for capacity building in the use of IPs

Both Bioversity and the CPWF have identified a need and begun training project staff and stakeholders on how to construct IPs. For the CPWF, this training also involves challenging researchers to become more entrepreneurial in spotting opportunities for the scaling-out and scaling-up of their research products. This culture shift includes expecting researchers to take a moral responsibility not just for the delivery of research products, but also to foster the formation of the networks needed for scaling-out and scaling-up. A number of CPWF researchers have used concepts and tools from their PIPA training in their own projects.[5]

[5] For example, see this document at *http://boru.pbwiki.com/f/Volta-MSC-stories.doc*

Issue 3: the role of IPs in monitoring and evaluation (M&E)

The existence of a clear impact theory helps to direct relevant data collection for M&E. Projects with explicit theories on how activities, outputs, outcomes and impacts appear to be linked are in a strong position to know what data are relevant and to collect those data in order to confirm or disconfirm the theories and so strengthen explanations of cause and effect. Monitoring, evaluating and updating IPs at specific project junctures constitute action research on the change processes initiated by a project. These activities generate new knowledge and understanding that feed back to the project and the networks in which it is embedded. This feedback contributes to adaptive management that responds to emerging opportunities and changing contexts. The issue remains as to how to achieve this feedback in ways that are practical for projects to implement.

Issue 4: the future role that the CGIAR SC and other leading promoters can play

In the CGIAR system, the SC often sets the standards for Center and Challenge Programme (CP) assessment and evaluation. The frequent focus of the SC has been *ex post* impact assessment through quantitative economics-based methods that calculate internal rates of return on research investment (Morris *et al.*, 2003). Priority assessment methods in general have received little attention. The SC and its Standing Panel on Priorities and Strategy have, by themselves, focused on determining System Priorities rather than on assisting and supporting the CGIAR Centers in their research planning efforts.

Ex ante assessment often cannot simply draw on methods employed in *ex post* analyses to date, which have principally focused on crop genetic improvement. The costs and benefits of many other types of R&D, particularly in more complex and multi-partner arrangements such as exist for natural resource management and in efforts to influence policy through research, are much harder to predict and calculate. In effect, such research is attempting to establish new IPs. Putting it another way, much germplasm research works in known contexts (plant breeders generally work with the same partners over many years) and with existing mechanisms (breeding in resistance to a pest is a known mechanism because it has already been done) and hence one can predict outcomes and impacts assuming history repeats itself. Policy research and natural resource management generally work in new contexts with new mechanisms and so predicting outcomes and impacts is much harder. Furthermore, they have a more evolutionary nature where objectives emerge in unpredictable ways as new information is gathered, and effectiveness and outcomes may be highly site-specific based upon local needs and adaptation processes. PIPA is a tool for surfacing and describing stakeholders' theories about these contexts (through network mapping), the mechanisms the project will use (through the outcomes logic models) and the expected outcomes and impacts. We have argued that this, in itself, is a form of priority assessment. If carried out in a flexible and iterative manner, it can help a project identify and manage evolving objectives and expectations.

To make further progress, it would be helpful if the SC and others who set standards for assessment and evaluation devote attention to defining good practice for priority assessment, and consider the potential contributions of PIPA and other approaches, such as Outcome Mapping (Earle *et al.*, 2001), in the process. Progress is being made in this direction. PIPA has been recommended as a precursor to *ex post* impact assessment in complex projects by the SC.[6]

Conclusions

There is growing interest in the use of IPs in the CGIAR system as a key element of the priority assessment and planning process. CGIAR Centers are innovating with different approaches to the construction and analysis of IPs. PIPA is one such promising methodology. Specifically, PIPA offers project managers and evaluators a practical set of tools that provide: (i) an appreciation of the existing and potential impact of research to set priorities and justify current and future funding; (ii) a deeper understanding of what impacts projects and programmes promise to attain as well as how and with which partners; and (iii) the framework for an effective M&E action research approach that fosters and tracks progress towards achieving impact. Hence, PIPA provides project managers with information for internal project priority setting in response to a changing environment and new opportunities and challenges. When PIPA is carried out across a number of projects, it allows programme managers (where a programme is made up of a number of projects) to prioritize between projects, based on both the desirability and the plausibility of the outcomes and impacts described in project IPs. M&E of progress along individual project IPs also provides programme managers with priority-setting information during programme implementation. Construction of programme IPs that show the added value of the programme also helps with priority setting because it clarifies how concrete project outputs contribute to wider programme goals, including the development of international public goods, as part of the CGIAR mandate. Clarifying this logic helps to identify gaps and funding priorities.

References

Bantilan, M.C.S. and Keatinge, J.D.H. (2007) Considerations for determining research priorities. Learning cycles and impact pathways. In: Loebenstein, G. and Thottappilly, G. (eds) *Agricultural Research Management*. Springer, 32.

Bennett, C.F. (1979) *Analyzing Impacts of Extension Programs*. United States Department of Agriculture, Washington, DC.

Campbell, B.M., Hagmann, J., Stroud, A., Thomas, R. and Wollenberg, E. (2006) *Navigating Amidst Complexity: Guide to Implementing Effective Research and Development to*

[6] Use of PIPA was recommended in a good practice guideline in a Science Council-commissioned publication (Walker *et al.* 2008).

Improve Livelihoods and the Environment. Center for International Forestry Research, Bogor, Indonesia.

CGIAR (2007) Guidelines for preparing CGIAR 2008–10 MTPs and 2008 Financing Plans. Available at: http://www.cgiar.org/pdf/finguide4_200701.pdf

Chen, H.T. (2005) *Practical Program Evaluation: Assessing and Improving Planning, Implementation, and Effectiveness*. Sage, Thousand Oaks, California.

CIFOR (1996) Annual Report. Available at: http://www.cifor.cgiar.org/publications/Html/AReport96.html

Douthwaite, B., Kuby, T., van de Fliert, E. and Schulz, E. (2003) Impact pathway evaluation: an approach for achieving and attributing impact in complex systems. *Agricultural Systems* 78, 243–265.

Douthwaite, B., Alvarez, S., Cook, S., Davies, R., George, P., Howell, J., Mackay, R. and Rubiano, J. (2007) Participatory impact pathways analysis: a practical application of program theory in research-for-development. *Canadian Journal of Program Evaluation* 22, 2 (Fall), 127–159.

Earl, S., Carden, F. and Smutylo, T. (2001) *Outcome Mapping: Building Learning and Reflection into Development Programs*. IDRC, Canada. Available at: http://www.idrc.ca/booktique

Ekboir, J. (2003) Why impact analysis should not be used for research evaluation and what the alternatives are. *Agricultural Systems* 78, 166–184.

Farrington, J., Thirtle, J. and Henderson, S. (1997) Methodologies for monitoring and evaluation in agricultural and natural resources research. *Agricultural Systems* 55(2), 273–300.

Fuglie, K. (2007) *Research Priority Assessment for the CIP 2005–2015 Strategic Plan: Projecting Impacts on Poverty, Employment, Health and Environment* International Potato Center (CIP), Lima, Peru. Available at: http://www.cipotato.org/publications/pdf/003744.pdf

Guijt, I. (1998) *Participatory Monitoring and Impact Assessment of Sustainable Agricultural Initiatives: An Introduction to Key Elements*. SARL Discussion Paper No 1, July. IIED, London.

Jones, A. and Seville, D. (2003) *Action-to-outcome Mapping: Testing Strategy with Systems Thinking*. The Systems Thinker, Pegasus Communications, Waltham, Massachusetts.

Kibel, B. (1999) *Outcome Engineering. Pacific Institute for Research and Evaluation (P.I.R.E.), Chapel Hill, North Carolina*. Unpublished manuscript.

Kuby, T. (1999) *Innovation as a Social Process: What Does This Mean for Impact Assessment in Agricultural Research?* Paper presented at a CIAT workshop, Costa Rica, September 1999.

Kuby, T. (2000) *Turning Attention Towards Results: How GTZ Is Building Its Impact Evaluation Capacity*. Internal GTZ document, Eschborn, Germany.

Mayne, J. (2001) Addressing attribution through contribution analysis: using performance measures sensibly. *Canadian Journal of Program Evaluation* 16(1), 1–24.

Mayne, J. (2008a) *Building an Evaluative Culture for Effective Evaluation and Results Management, ILAC Brief No. 20*. Institutional Learning and Change (ILAC) Initiative, Rome, Italy.

Mayne, J. (2008b) *Contribution Analysis: An Approach to Exploring Cause and Effect, ILAC Brief No. 16*. Institutional Learning and Change (ILAC) Initiative, Rome, Italy.

Mayne, J. (forthcoming) Contribution analysis: addressing cause and effect. In: Schwartz, R., Forss, K. and Marra, M. (eds) *Evaluating the Complex*. Transaction Publishers, Edison, New Jersey.

Morris, M., Pingali, P., Gregersen, H. and Kelly, T. (2003). Assessing the impact of agricultural research: an overview. *Quarterly Journal of International Agriculture*, 42(2), 127–148.

Pender, J., Place, J. and Ehui, S. (1999) *Strategies for Sustainable Agricultural Development in the East African Highlands*. EPTD Discussion Papers No. 41. IFPRI, Washington, DC.

Rogers, P. (2008) Using programme theory to evaluate complicated and complex aspects of interventions. *Evaluation* 14, 29–48.
Rossi, P.H., Lipsey, M.W. and Freeman, H.E. (2003) *Evaluation: A Systematic Approach*, 7th edn. Sage, Thousand Oaks, California.
Springer-Heinze, A., Hartwich, F., Simon Henderson, J., Horton, D. and Minde, I. (2003) Impact pathway analysis: an approach to strengthening the impact orientation of agricultural research. *Agricultural Systems* 78, 267–285.
Suchman, E. (1962) *Evaluation Research: Principles and Practice in Public Service and Social Action Programs*. Russell Sage Foundation, New York.
Walker, T. and Collion, M. (1997) *Priority Setting for CIP for the 1998–2000 Medium Term Plan*. International Potato Center (CIP), Lima, Peru.
Walker, T., Maredia, M., Kelley, T., La Rovere, R., Templeton, D., Thiele, G. and Douthwaite, B. (2008) *Strategic Guidance for Ex-Post Impact Assessment of Agricultural Research*. Prepared for the Standing Panel on Impact Assessment, CIGAR Science Council, Rome, Italy. Available at: http://www.sciencecouncil.cgiar.org/fileadmin/user_upload/science-council/Impact_Assessment/SC_epIA_low-res_for_web.pdf
Weiss, C. (1972) *Evaluation Research: Methods for Assessing Program Effectiveness*. Prentice-Hall, Englewood Cliffs, New Jersey.
Weiss, C.H. (1995) Chapter 3 The haphazard connection: social science and public policy. *International Journal of Educational Research* 23(2), 137–150 Available at: http://www.sciencedirect.com/science/article/B6VDF-3YCMV9Y-1B/2/17cf1edfceca6f530c85ad6c48e59f50
Wholey, J.S. (1977) Evaluability assessment. In: Rutman, L. (ed.) *Planning Useful Evaluations: Evaluability Assessment*. Sage, Beverly Hills, California.
Yampolskaya, S., Nesman, T.M., Hernandez, M. and Koch, D. (2004) Using concept mapping to develop a logic model and articulate a program theory: a case example. *American Journal of Evaluation* 25(2), 191–207.

Appendix 2.1

Using an impact pathways approach to identify priority interventions at Center, discipline and CP levels

Classification: Management and Coordination

Name of Person Reporting: Malcolm Beveridge (Theme 3 Management Team)

Project/Theme/Basin: CPWF Theme 3

Date when the change occurred: September 2006

Place where the change occurred: Throughout all areas of work related to aquaculture

The Story:

The WorldFish Center promotes the use of aquaculture to tackle poverty. Specifically, it uses aquaculture to directly and indirectly make substantial contributions to the development of sustainable, resilient livelihoods, create employment and stimulate economic growth and improve the nutrition of many millions of poor people. However, while aquaculture has fulfilled its promise in many parts of Asia, in sub-Saharan Africa its successes, like those of attempts to increase agricultural output, have been mixed. As a result of our experiences working with the CP on Water and Food, we decided that using an Impact Pathways approach to identify priority interventions could prove useful in showing where we should focus our research efforts and on how we build partnerships to scale up and out impacts from research projects.

The Problem Tree for aquaculture asserts that the key problem is that aquaculture cannot substantially help poor people escape from poverty, and employs an IP approach to tease out the causal links. By stating the reverse – i.e. by asserting that aquaculture substantially helps poor people escape from poverty – and by mapping the causal links we can identify the types of broad, high impact interventions required to effect the uptake of aquaculture as a cost-effective means of addressing poverty. The Problem Tree helped identify four key thematic areas:

Aquaculture as an engine for rural economic development. Poor people often have the resources and skills to use aquaculture to help lift themselves out of poverty. However, there can also be barriers to the adoption of aquaculture by the poor, including poor education and health, access to resources and input and output markets, technological knowledge and management skills, a poor enabling environment and political and social constraints.

Development and dissemination of sustainable aquaculture technologies. Aquaculture is still in its infancy. Many emergent new technologies promote better use of resources and reduced demands on environmental services, increased productivity and production, and result in cheaper products. Efforts are needed to ensure greater access to these technologies, especially quality seed and feeds, constraints to aquaculture fulfilling its potential to help people escape from poverty, that we believe are as much the result of inefficient and ineffective private/public sector interactions as technological impediments.

Protection and enhancement of environmental quality. Aquaculture must be implemented in as environmentally sound a manner as possible. Adoption of aquaculture technologies can lead to better use of resources (increased water productivity, recycling of on-farm materials). However, over-expansion of aquaculture makes unsustainable demands on environmental goods and services, with the poor proving to be most vulnerable. To substantially contribute to reducing poverty, aquaculture must be incorporated into watershed and coastal management plans. Greater understanding of the risks associated with intensification (disease, self-pollution), translocation of species and genetically improved strains, in adoption of cage aquaculture and in exploitation of environments vulnerable to climate change, is needed.

Development of human and institutional capacity. The successful and sustained adoption of aquaculture to impact on poverty reduction requires development of capacity among policy makers and national research and extension institutions and staff. This empowerment requires identification of skills needs, training requirements and professional development support, opportunities to strengthen the capacity of institutions to deliver training and continuing professional development through a range of models, including distance learning, and identification of opportunities for scientific mentorship among institutions (e.g. networks).

Why is the story significant?

We believe that this type of analysis clearly presents a comprehensive view of the intervention logic, explains how project activities and outputs will contribute to a sequence of outcomes and impacts and facilitates constructive discussion among project team leaders. The importance of most significant change stories, in helping clarify and communicate the research for development processes out of which impact emerges, also became clear (see Douthwaite *et al.*, 2003, http://impactpathways.pbwiki.com). For me, as a member of the CPWF T3 Management Team, it also helped ensure that the right priority research areas had been identified and that complementary issues were being pursued within our Center.

What were the critical factors that led to the change?

It was my engagement with the CPWF some six months after joining the WorldFish Center that opened my eyes to this exciting way of analysing problems that went far beyond the more widely used logistical framework approach in trying to ensure development impact. The International Forum on Water and Food in Vientiane later that year provided further opportunities to develop my thinking and to work with others to try to use the impact pathway and most significant change story approach to achieving development impact.

What were the constraints?

Time.

What are the future implications for actions (e.g. future research), if any?

Within the Center, we intend to use models as a starting point for discussions with staff and partners at a regional level, allowing further refinement and crystallization of priority interventions. A process of strategising is then needed to decide specifically in which arenas WorldFish might lead, where it might partner and where it will merely champion actions.

3 Research Priority Assessment at the International Potato Center (CIP)

KEITH O. FUGLIE AND GRAHAM THIELE*

Abstract

Since the 1990s, CIP has employed formal economic *ex ante* and *ex post* assessment of research to help guide research resource allocation at the Center. Initial work in the mid-1990s on *ex ante* priority assessment developed estimates of anticipated impacts for each constraints-based project in the Center's research programmes on potato, sweet potato and natural resource management. These estimates confirmed that the Center's investment in most of these projects was broadly appropriate and contributed to some realignment of projects that appeared to have lower impact potential. Subsequent *ex post* impact studies revealed that some of the research projects were unlikely to achieve their anticipated impacts, whereas other projects exceeded their projected impacts. A second round of programme-wide *ex ante* priority assessment was carried out in 2005 as part of the Center's strategic plan. The findings from the *ex post* impact studies were an important input to the new assessment exercise and contributed substantially to revised estimates of anticipated impacts. This impact assessment went beyond economic surplus measures to model the likely impacts of research on metrics for poverty. The influence of formal priority assessment on Center programmes diminished, however, as increased reliance on restricted funding reduced the Center's flexibility in allocating research resources.

Keywords: Priority assessment, potato (*Solanum tuberosum*), sweet potato (*Ipomoea batatas*), economic surplus, poverty impact, returns to research, international agricultural research

* Keith Fuglie was the leader of CIP's Impact Enhancement Division during 2003–2006 and is now with the U.S. Department of Agriculture's Economic Research Service (ERS) in Washington, DC. Graham Thiele currently serves as leader of this division. However, the views expressed in this chapter are the authors' own and do not necessarily reflect the position of CIP or ERS. The authors are solely responsible for any errors or omissions.

Introduction

The International Potato Center (CIP) was established as an institute of the Consultative Group on International Agricultural Research (CGIAR) in 1973 with a mandate to improve potato productivity in developing countries. In 1985, the research mandate was expanded to include sweet potato. In the early 1990s, CIP added programmes in natural resource management in mountain areas (with an emphasis on the Andean region) and germplasm conservation and crop improvement programmes for other minor Andean root and tuber crops. CIP's efforts to improve agricultural productivity in developing countries broadened from raising crop yield to increasing crop value through postharvest research and strengthening natural resource management in agricultural systems, especially in the Andean region.

During the 1980s, efforts to establish research priorities at CIP were based on 'planning conferences' in which crop scientists from CIP, developing countries and some developed country institutions met to identify key productivity constraints and discuss possible research approaches to address them (e.g. CIP, 1984). Occasionally, formal surveys of scientists would be conducted to elicit their opinions or rankings on priority issues for commodity improvement (Horton, 1989; Herrera and Scott, 1992/93). These exercises helped to build consensus around important issues for CIP's research programme during a period when Center resources were expanding. But they did not provide much information on the relative impact of research projects or the trade-offs from investing in some kinds of research versus others.

In the 1990s, the need to make hard choices among alternative research strategies became more apparent when CIP faced stagnating or declining research resources. Around this time, CIP began to invest in formal priority assessment as well as impact assessment. CIP's history of research priority assessment is summarized in Table 3.1. The first formal priority setting exercise at CIP was conducted in 1991 and was jointly led by CIP's Deputy Director General for Research Peter Gregory and Marie-Hélène Collion of International Service for National Agricultural Research (ISNAR). They used a scoring model developed by ISNAR in which each constraint-releasing research project at CIP was evaluated against a number of criteria by CIP scientists (Collion and Gregory, 1993). In 1995, Thomas Walker, then Head of Social Sciences at CIP, and Marie-Hélène Collion conducted a second priority-setting exercise with a number of refinements: They improved on the first exercise by: (i) redefining the research portfolio to draw a clearer link between research and impact; (ii) introducing benefit–cost analysis to rank projects (using net present value (NPV) and internal rate of return (IRR)); (iii) weighting anticipated research benefits by the extent of technology adoption in poor areas; and (iv) deriving results by country rather than by agro-ecological regions, which was more useful for planning resource allocation (Walker and Collion, 1997). They used the estimates of the relative returns to the various projects to make suggestions for programme realignment.

Subsequent to the 1995 *ex ante* priority assessment, CIP completed more than a dozen *ex post* impact assessments of CIP-related technologies (Walker and Fuglie, 2001). These case studies focused primarily on documenting the

Table 3.1. CIP's research priority assessments. (From authors' estimates.)

	1991	1995	2005
Method	Scoring	Discounted net value of probable adoption by country	Discounted net value plus, identification of impacts on producers and consumers based on estimates of poverty-weighted producer and consumer welfare and human health
Gain in precision from prior exercise	Quantitative rankings of research projects	Anticipated benefits related to cost of research Benefits weighted for poverty impact, but this did not affect project rankings	Closer attention to modelling poverty impacts indicated higher priority should be assigned to sweet potato
Approximate cost in IRS person days	100	200	370[a]
Effects on Center management	Center divested of some low-return research (nematode)	Confirmed Center's main priorities Guided *ex post* assessment, which led to divestment of some low-return research areas (TPS, sweet potato product development)	Too early to judge

[a] Senior economist – 220 days in planning, collecting and analysing data and write up; science leaders and regionally posted staff – 150 days during three different consultations.

impact of the Center's 'success' stories but also included reassessments of technologies where adoption rates appeared to be substantially less than anticipated in the 1995 priority assessment. The results of the case studies influenced opinions among scientists at the Center on the likely impacts of several of the projects in CIP's research portfolio.

Then, in 2005, CIP undertook another Center-wide priority assessment of its research programme as part of a strategic planning exercise. The exercise was led by Keith Fuglie, the leader of CIP's Impact Enhancement Division at that time, and involved all CIP scientists at its main station in Peru as well as out-posted scientists. This round of priority assessment took as a starting point the method and results of the 1995 exercise. The new assessment process was also able to draw upon the lessons learnt from the *ex post* impact studies, and consequently, the anticipated impacts of many of the Center research endeavours were significantly revised. It also had more in-depth analysis of welfare and poverty consequences, and included quantitative assessments as well as qualitative scoring models of poverty, health, environmental and labour impacts of research (Fuglie, 2007). This chapter describes the priority assessments carried out at CIP, focusing on the 1995 and 2005 exercises. We briefly describe the methods, key findings, some limitations of the analysis and how this information may have influenced decision making at the Center.

Methods and Key Findings

In this section, we describe the methods used for quantitative assessment of research priorities at CIP and some of the key results of the exercise.[1] We begin by presenting the framework developed by Collion and Gregory (1993) and Walker and Collion (1997). We then show how *ex post* impact assessments contributed to institutional understanding of prospects for CIP-related technologies. Finally, we describe how Fuglie (2007) extended this framework in CIP's most recent priority assessment exercise.

Project appraisal framework for *ex ante* impact assessment

The framework developed by Collion and Gregory (1993) and revised by Walker and Collion (1997) for assessing research priorities required a number of essential steps. First, a target adoption area was defined. The target area is the domain over which potential research impact was assessed. It consisted of a subset of prioritized developing countries selected on the basis of average income level, poverty rate and the importance of potato and/or sweet potato in the country (CIP, 2004). Each country in the target area was assigned to one or more agro-ecological zones for potato and for sweet potato based on environmental conditions.

Second, a set of constraints-based technologies ('projects') was identified for the analysis. Attention was given primarily to CIP's existing research portfolio but also included input from CIP stakeholders on new or emerging issues of potentially high importance. Existing or potential research activities were grouped into reasonably self-contained research endeavours or 'projects' that targeted a well-defined constraint or opportunity to achieve practical impact. By 'self-contained' research project we mean a set of research activities that

[1] Although we included a number of qualitative impact indicators in the technology assessment exercise, our focus in this discussion is on the quantitative part of the assessment. The quantitative results gave sharper indications of trade-offs among alternative research investments (Fuglie, 2007).

does not depend directly on the results of another research project, and produces outputs which can be taken up by extension and development organizations and which has a well-defined impact target. Self-contained projects combine some basic research activities together with more applied activities that are directed towards the same outcome. For example, a project on reducing potato yield constraints due to late blight disease might include activities on molecular studies of plant–pathogen interaction, identification of novel genes for resistance, breeding locally adapted varieties and development and adaptation of integrated control methods in the field. However, a research activity on 'integration and adaptation of technology' is likely to be dependent on upstream research to develop the technology, and thus would not be considered a self-contained project. Some Center activities, such as crop germplasm conservation and impact assessment itself, were considered as Center service responsibilities and were not included in the priority assessment.[2]

The third step was to elicit judgements from scientists on the potential to advance technology for reducing a productivity constraint in specific agro-ecological zones, assuming a given amount of research expenditure over a fixed period of time. Teams of scientists reached group consensus of technical assessments for specific projects for each agro-ecological zone where the productivity constraint was thought to be significant. This technical assessment included judgements from scientists on the likelihood of the project to improve crop yield or value, change production costs and on the likely adoption rate. Once estimates were formed for agro-ecological zones, revisions were made for individual countries in a zone based on factors specific to that country. Potential adoption of the technologies in non-target countries was identified as a 'spillover effect' of the research and did not factor directly into the priority assessment.[3] The final step was to combine these data with models to derive expected benefits and their relation to the investment in research.

The basic model used by Walker and Collion (1997) (as well as by Collion and Gregory, 1993) for deriving total expected benefits from research on a particular technology, once full adoption is achieved, was:

$$E(B) = \left[\frac{E(B)}{Ha}\right] * [Ha] * [\mathrm{Pr}], \tag{3.1}$$

where $E(B)$ = Total expected benefits at full adoption;

$\left[\frac{E(B)}{Ha}\right]$ = Expected benefits per hectare of adoption, including the value of quality and yield improvement and cost changes;

[Ha] = Adoption ceiling, in hectares;

[2] Although not included in the Center's priority assessment, some service activities like germplasm conservation have been subjected to economic impact evaluation at the CGIAR-systems level (Koo *et al.*, 2004).

[3] Spillover countries were non-target countries with similar agro-ecology that would likely benefit from the research. In this context, target countries can be thought of as the primary recommendation domain and spillover countries as the secondary recommendation domain.

[Pr] = Probability that the research project would be successful in achieving these outcomes.

Walker and Collion (1997) added benefit–cost analysis to derive estimates of NPV and IRR for each research project. These were estimated by forming an annual stream of research costs and adoption benefits. To get an annual stream of expected benefits, technology adoption was assumed to occur along a logistic diffusion curve to reach the adoption ceiling (with sensitivity analysis for slow and rapid rates of adoption). Estimates of research impact where adoption was expected to take place were provided for each technology in each country, or for large countries, for each province or state. Impacts were aggregated to get global results as well as results by region, agro-ecological zone and commodity. The expected benefits were also weighted for their 'poverty content' by multiplying the expected benefits in a country by the World Bank's US$1/capita/day headcount poverty index, estimated at purchasing power parity, for that country. This procedure gives greater weight to research that addresses productivity constraints that are more important in poor areas, although it assumes that within these areas benefits would be evenly shared among income groups.

The results of the 1991 and 1995 priority assessments confirmed what had already been established as the principal priorities for CIP research: addressing a number of key biotic constraints of potato and sweet potato as well as enhancing dry matter yield and market utilization of sweet potato. The exercises did, however, provide further justification for the institution to divest of a few activities, namely research on nematodes and abiotic constraints, which ranked low in the priority assessment, as well as remove middle-income countries in Latin America and North Africa/Middle East from the CIP priority target area (Walker and Collion, 1997).

Testing the assumptions with *ex post* impact case studies

Ex post case studies of technology impact conducted by CIP in the 1990s largely followed the project appraisal framework described above, except that actual outcomes were used instead of anticipated results. *Ex post* impact assessment at CIP was shaped by a felt need to document the adoption and consequences of CIP-related production and postharvest technologies, but it also served to inform the priority-setting process. Thus, the emphasis was on case studies of success stories. By 2001, 12 case studies had been elaborated in the same format and published in a 'red and black' impact series (see Table 3.2). An aggregate spreadsheet tallied net benefits of the success stories against CIP's total cost since its founding in 1971. While the 12 case studies did not represent the entirety of CIP's impact, the aggregated tally provided evidence of solid returns to donor investment in the Center as a whole (Fig. 3.1).

One limitation of these 'success stories' is that no attempt was made to allocate total benefits among CIP and its partners, although counterpart

Table 3.2. Case studies of *ex post* impact of CIP-related technologies. (From Walker and Fuglie, 2001.)

Technology: General	Technology: Specific	Country	Timespan for project appraisal	Returns on investment[a]: IRR (%)	Returns on investment[a]: NPV (US$ million)
Varietal	Potato – late blight resistance and improved seed supply	Rwanda Burundi Zaire	1978–1993	92	27
	Potato – resistance to drought and viruses	China	1978–2000	106	11.9
	Potato – late blight resistance	Peru	1979–2020	27	5.4
	Sweet potato HYV	Peru	1990–2020	44	3.0
Integrated pest management	Potato tuber moth	Tunisia	1976–2000	64	6.4
	Sweet potato weevil	Dominican Republic	1989–2019	29	1.1
	Sweet potato weevil	Cuba	1993–2020	65	21.7
	Andean potato weevil	Peru	1988–2018	32	1.8
Seed	Rapid multiplication and late blight resistant varieties	Vietnam	1978–1993	81	2.1
	True potato seed	India	1978–2015	29	18
	Sweet potato virus free seed	China	1978–2015	202	550
	True potato seed	Egypt	1979–2015	28	2.9
	True potato seed	Vietnam	1990–2010	39	1.8

[a] IRR = internal rate of return; NPV estimated using a 10% discount rate at 1996 prices.

research and extension costs by National Agricultural Research Systems (NARS) were netted out of the benefits estimation. Thus, the total net benefit in Fig. 3.1 is actually a return to combined CIP–NARS investments and not to CIP alone. None the less, each case study did identify a critical scientific or technical contribution by CIP that was necessary for the success of the project. One impact study, on postharvest utilization of sweet potato in China, found evidence of significant economic impacts but could not firmly attribute much of this to CIP and thus was not counted in the tally of CIP impacts (Fuglie, 2006).

The attention given to *ex post* coverage of success stories did not mean, however, that 'failures' were not analysed. Lack of evidence of technology adoption triggered studies to assess adoption constraints. Anticipated impacts were subsequently downgraded for a number of projects, including research on

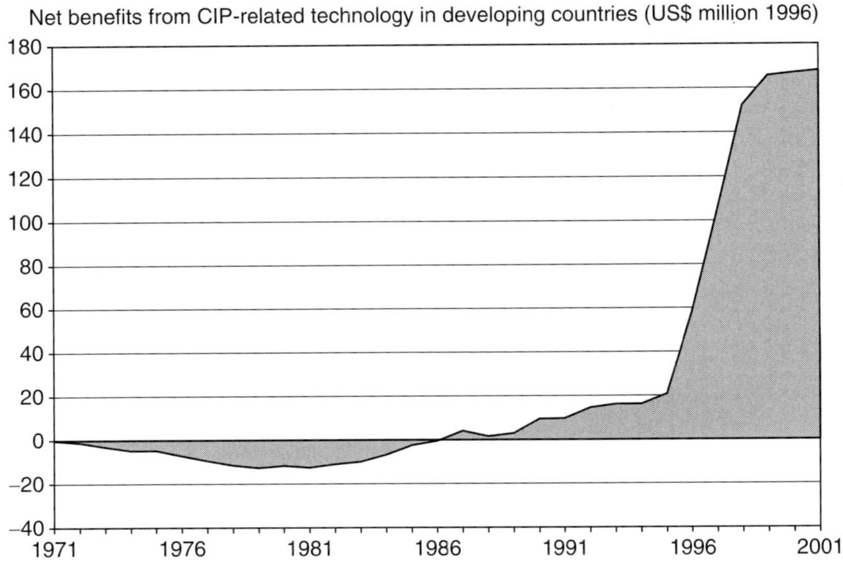

Fig. 3.1. Total annual net benefit from investing in CIP documented through impact assessment (millions of US dollars at constant 1996 prices; Walker and Crissman, 1996).

rustic potato storage (Fuglie *et al.*, 2000), true potato seed[4] (Chilver *et al.*, 2005) and product development for root and tuber crops (Walker and Fuglie, 2006). These assessments highlighted areas for improvement in technology design or for divestment until conditions would warrant further investment. A comprehensive review of impact in Asia conducted by Fuglie (2006) showed that in most cases the 1995 projections of technology adoption were too optimistic. While in one case the documented adoption exceeded project expectations, by 2004 overall adoption of the set of technologies in CIP's portfolio in Asia was only about one-fourth the area anticipated by 2015 (Table 3.3). Many of the projects were unlikely to see much further impact without significant improvements to the technology.

Deepening the analytical framework for priority assessment

A limitation of the modelling framework described above for impact assessment is that it relies heavily on a single impact criterion – total economic surplus. While the 1995 priority assessment did weigh benefits by a poverty headcount index, it gave no further consideration to the distributional consequences of technological change. It was basically assumed that everyone in a country would share about equally in the economic benefits generated in that country.

[4] This is a botanical seed as opposed to the tubers which are normally planted as 'seed' potato.

Table 3.3. Adoption of new potato and sweet potato technologies in Asia. (From Fuglie, 2006.)

Technology	1995 forecast of adoption level in 2015 ('000 ha)	Adoption achieved by 2004 ('000 ha)	Are forecasts likely to be reached?
Potato technologies			
Late blight control	644	200	Yes, if dissemination of new varieties can be speeded up
True potato seed	135	15	No. Economic assessments have shown that applications of TPS will be more selective
Virus control and improved seed systems	345	40	Yes, if dissemination of new varieties can be speeded up
Bacterial wilt control	125	–	No
Integrated pest management	143	–	No. Dissemination of IPM technology has proven to be complex and costly
Improved cropping systems	39	5	Yes. Adoption of double rice transplanting in potato–rice system is beginning to take off
Improved potato storage	135	–	No. Economic assessment of storage technology showed that adoption is unlikely
Sweet potato technologies			
High starch-yielding varieties	1633	10	Yes. Improved varieties just now becoming available
Improved postharvest utilization	844 (all)	10	1. Starch processing: No, private sector has become dominant source of new technology
		15	2. Animal feed: Unlikely, unless policies give more attention to backyard pig production
Virus control and improved seed systems	573	800	Yes, have already exceeded expectations
Integrated pest management	373	5	No. Dissemination of IPM technology has proven to be complex and costly
	4989	1100	

The distributional consequences of technological change could exert a significant influence on poverty impact, an important consideration for the CGIAR. New technologies that shift supply or demand for commodities are likely to affect market prices, and these price effects will influence the distribution of benefits among producer and consumer groups. The market price effects are likely to be important for commodities like root and tuber crops that are not widely traded, so that prices are determined by local supply and demand conditions. The pattern of commodity utilization will also affect welfare distribution as subsistence farmers will benefit both as producers and consumers of the increased output (Herdt and Hayami, 1977). The distributional consequences are likely to be of greater interest in cases where producers and consumers of a commodity fall into different socio-economic classes.

The potential downward pressure on market prices (and farm income from crop sales) as adoption of new technology increases commodity supply has been a concern in the potato and sweet potato research communities. It has been of particular concern where per capita consumption of a commodity declines (or grows only slowly) with increases in per capita income. This appears to be the case with sweet potato in much of Asia and Africa and potato in the Andes region. These concerns have been given expression by the research and development community through interest in developing technologies to create new uses and markets for these crops, such as breeding varieties suitable for making processed products and improving efficiency of animal feed utilization. What is probably less well-understood among scientists and research managers, however, is that if research causes market prices to fall, it does not mean that the value of the research is falling. The primary effect of falling prices resulting from technology adoption is to shift a share of the economic impact from producers to consumers. The share of benefits from research going towards poverty reduction will depend upon the extent of poverty among both the producers and consumers of the commodity. If the commodity factors largely in the diets of the very poor, research that results in lower food prices could have a very favourable impact on poverty, even though some producers (i.e. non-adopters who sell some output to the market) may experience a fall in income.

In the 2005 priority assessment exercise, Fuglie (2007) explicitly accounted for the price effects of technical change and the distribution of economic benefits among producer and consumer groups. The basic model is shown in Fig. 3.2. In the figure, the effect of technological change is modelled as a downward shift in the supply function, reflecting a lower unit cost of production when yield increases or input use is reduced. The estimate of total economic surplus in Fig. 3.2 is similar to the total expected benefits given in Eq. 3.1,[5] but the added

[5] Equation 3.1 is equivalent to the changes in total economic surplus shown by the shaded regions in the special case where the supply elasticity is 1 and the demand elasticity is 0 (a supply elasticity of 1 assures that percentage changes in yield are equivalent to percentage changes in the marginal cost of production, and a demand elasticity of 0 will leave the market price unchanged even as market supply is increased). For other elasticity values the total surplus measures are very similar – alternative assumptions about supply and demand mainly affect the distribution of total surplus among producers and consumers and not the measure of total surplus (Alston, Norton and Pardey, 1995).

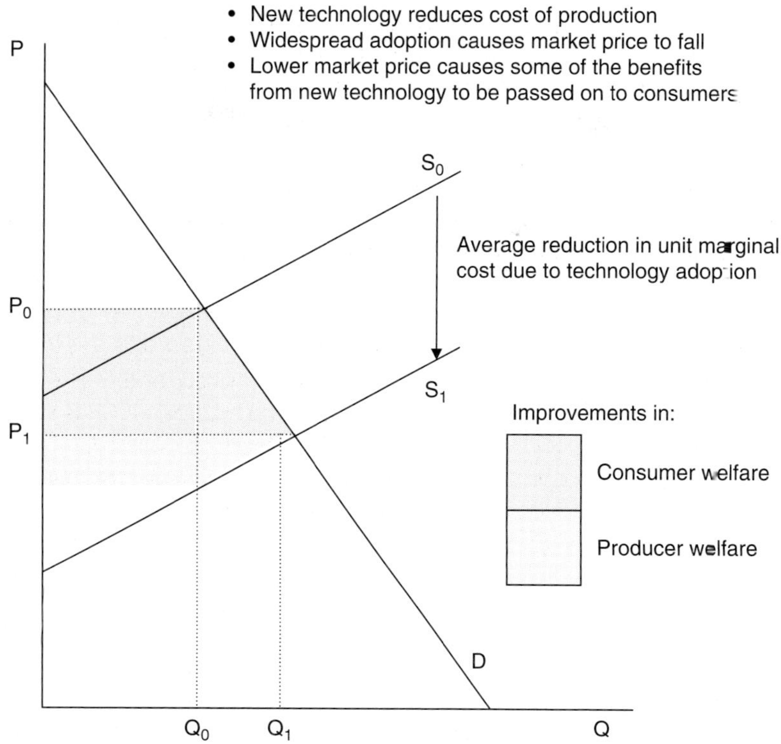

Fig. 3.2. Distribution of welfare benefits from technological change that lowers production cost.

value of the figure is that it shows the distribution of benefits going to producers and consumers of the commodity. The model in Fig. 3.2 was adapted to semi-subsistence agriculture (where a significant part of output is used on-farm as either food or feed) to derive impacts on income of adopting and non-adopting households. Other models were developed to evaluate technologies that improve postharvest utilization or expand market demand for a commodity (incorporating shifts in the demand curve rather than the supply curve). In addition to economic surplus measures, qualitative scoring models and the disability-adjusted life years (DALYs) method was used to evaluate impacts of technologies on human health. With estimates of per capita impacts for producer and consumer groups and using World Bank data on income distributions, estimates could be derived for the number of persons likely to be lifted out of poverty.

Estimating the distributional consequences and measuring impacts on health and poverty required considerably more socio-economic information and modelling effort compared to the approach described in Eq. 3.1 and with some loss of transparency to the process. Isolation of the likely income and poverty impacts (which may not be well measured by either consumer or producer surplus) required information on supply and demand elasticities, the share of output used on the farm, the share of marketed surplus consumed by poor non-farm households, the size distribution of farms, factor biases in the technology, the likely diffusion profiles for different socio-economic farm classes and income distributions in affected countries. Estimation of the number of DALYs saved from improved nutrition required information on the health status of different consumer groups and how changing the micronutrient content of food crops would likely affect their health outcomes.[6] Microeconomic data on potato and sweet potato-producing households were drawn principally from farm surveys conducted by CIP social scientists since the 1970s. Estimates of demand elasticities for potato and sweet potato in developing countries were developed from an extensive review of empirical studies, although supply elasticities were largely assumed. FAO production data, WHO health data and World Bank data on income distributions were other important sources of information. See Fuglie (2007) for a detailed description of these data sources and the models.

Some Key Findings

The results of CIP's priority assessment exercises provided quantitative indicators of the likely pay-offs from the various components of CIP's research portfolio. Table 3.4 presents the summary benefit–cost analysis of CIP's principal research projects from the 2005 exercise. One of the striking features is how large the expected benefits are relative to the cost of research. The global NPV of CIP's potato and sweet potato research (using a 10% real discount rate) over the 2006–2030 period ranged from US$1.65 billion to US$3.07 billion, depending on the adoption scenario. High returns are not unusual for *ex ante* assessment exercises even when scaled by probabilities of success. One factor may be general optimism on the part of scientists on prospects for research success and technology diffusion. Scientists' estimates of the likelihood for success of the technologies listed in Table 3.4 ranged from 0.52 to 0.90. Scientists may also overstate the potential (or under-appreciate the cost) of cross-national technology adaptation and spillover – measured by the estimated adoption ceilings for various countries and regions. A second feature of the results reported in Table 3.4 is the wide variation in returns to research across technologies. Internal rates of return (under the slow adoption scenario for aggregate benefits) ranged from negative to +56%. Generally, breeding and improvement of clonal seed propagation gave the highest returns for both potato and sweet potato, while investments in pest and disease management, cropping systems and postharvest market development gave significantly lower returns. The returns to research are highly correlated with estimates of adoption area, which is consistent with the results of

[6] The health impact model was a modified version of the model proposed by the *HarvestPlus* Challenge Programme (Stein *et al.* 2005). Some assumptions used in the *HarvestPlus* model, however, appear to give rather generous estimates of human health benefits from improved nutrition. Our modified model gives a more conservative assessment of the health impact of reducing micronutrient deficiency (see Fuglie and Yanggen, 2006).

Table 3.4. Benefit–cost analysis of CIP potato and sweet potato research, 2006–2030. (From Fuglie, 2007.)

	Adoption	Aggregate benefits (US$ '000)				Benefits to rural poor (US$ '000)			
	Ceiling	Rapid (2020)		Slow (2030)		Rapid (2020)		Slow (2030)	
Technology	(ha)	NPV	IRR (%)	NPV	IRR (%)	NPV	IRR (%)	NPV	IRR (%)
Potato late blight breeding and management	693,212	889,805	57	482,998	39	104,733	25	40,868	16
Potato pathogen-free planting material	653,990	501,993	59	272,794	41	52,350	25	19,568	16
Potato breeding for virus resistance	220,450	336,486	52	184,714	38	32,372	23	13,445	16
Potato breeding for processing	154,890	(196)	9	(1,378)	5	25,027	37	6714	21
Potato bacterial wilt management	100,690	63,878	38	32,657	26	2184	13	(2087)	7
Potato botanical seed propagation (TPS)	44,063	29,977	32	14,927	23	2521	14	(535)	9
Potato IPM of insect pests	129,244	63,776	31	30,754	22	219	10	(5039)	6
Potato market chain improvement	33,421	5035	16	173	10	(1292)	8	(3390)	4
Potato–cereal cropping systems management	40,271	17,143	42	8732	28	404	12	(695)	6
Total potato	2,070,232	1,907,896	49	1,026,369	35	218,518	22	68,847	14
Sweet potato pathogen-free planting material	907,577	803,924	82	443,107	56	214,336	49	110,269	31
Sweet potato breeding for high Vitamin A	139,626	133,618	41	70,214	29	69,895	32	34,241	22
Sweet potato breeding for high dry matter	314,791	62,730	38	24,653	25	26,646	28	11,056	19
Sweet potato IPM for weevil control	197,332	103,877	58	54,989	37	57,877	45	29,020	28
Sweet potato utilization for animal feed	62,088	65,932	38	34,266	27	6757	17	861	11
Sweet potato utilization for small enterprises and new products	77,258	(3687)	<0	(3722)	<0	1346	14	(881)	8
Total sweet potato	1,698,672	1,166,395	56	623,508	38	376,857	37	184,566	24

NPV calculated using a 10% real discount rate.
Ex ante benefits are estimated assuming a 'rapid' scenario in which the adoption ceiling is reached by 2020 and a 'slow' adoption scenario in which the ceiling is reached by 2030.

the 1995 exercise. What Walker and Collion concluded in 1997 held equally true a decade later:

> Summing up, *area coverage* strongly conditions the results. It sounds trivial, but still warrants stating: CIP should place its emphasis not on the most severe per-unit constraints but *on the most widespread ones*.
>
> (Walker and Collion, 1997, p. 18, *italics added*)

A few other results from the 2005 priority assessment exercise are worth drawing attention to. One is that sweet potato registered higher impacts on poverty than potato. The additional effort put into modelling poverty impacts in the 2005 exercise resulted in a higher poverty profile for this crop. The result follows from the role of sweet potato as an important subsistence, nutritional and food security crop in some of the poorest regions of the world, especially in sub-Saharan Africa, even though in terms of price sweet potato is a relatively low-value crop. A second result worth highlighting is that anticipated impacts were geographically concentrated (Table 3.5). North-east Asia (China, primarily) accounted for 63% of the total anticipated economic impact and 41% of impact on poor households. These numbers correlate closely with total crop area and anticipated adoption rates: China alone accounts for about 70% of the combined crop area of potato and sweet potato in developing countries and has a strong extension system to disseminate technology. Sub-Saharan Africa accounted for only 17% of aggregate impact but 44% of poverty impact due to the greater prevalence and depth of poverty in this part of the world, and the importance of these crops in food security and nutrition of the rural poor. Both these results pointed to imbalances in CIP's allocation of research resources in 2005, which favoured potato over sweet potato by a 3:1 margin even though poverty impacts were balanced between the commodities. Similarly, about 40% of the Center research resources were focused on achieving impact in the Andes countries even though only 4% of the potential poverty impacts were anticipated from this region.

Finally, a further finding which deserves comment is the relative impacts of production versus post-production research. In the 1995 exercise, high returns were anticipated for research to develop new processed food products from sweet potato. Subsequently, it became clear that where large impacts of processing and product development were taking place (namely in China), the private sector was by far the dominant player. As a result, in the 2005 exercise, scientists significantly scaled back the anticipated benefits that could be attributed to CIP research on post-production sweet potato compared with their expectations from a decade earlier. However, large poverty impacts were now anticipated from breeding potato varieties suitable for processing. This impact stems from the positive effect on commodity prices due to the growing demand for these products. Here, higher prices from technological change served to shift benefits from relatively well-off consumers to relatively poor producers. But the estimates of these benefits are sensitive to assumptions about attribution, i.e. how much of the growth in demand can reasonably be attributed to

Table 3.5. Regional distribution of anticipated impacts of potato and sweet potato research in 2030 (measured in US$ '000/year in aggregate benefits after adoption of ceiling reached). (From Fuglie, 2007.)

Region	Adoption area (ha)	Aggregate impact	Benefits to rural poor	Percentage of global aggregate impact	Percentage of global benefits to rural poor
Sub-Saharan Africa	982,877	194,116	118,006	17	44
North-east Asia	1,966,092	723,380	109,020	63	41
South and Central Asia	505,850	160,696	26,961	14	10
Latin America	211,753	68,403	10,624	6	4
South-east Asia	102,332	28,499	2,725	2	1
Global	2,070,232	1,146,594	267,336	100	100

CIP research and how much would occur anyway, led by the private sector and general economic growth. The large reduction in expected benefits of CIP's post-production research for sweet potato between the 1995 and 2005 exercises followed from a realization that CIP-related technologies were being supplanted by technologies from other sources, especially from the private sector. That may turn out to hold true for potato as well.

Discussion

The influence of quantitative assessment of research priorities at CIP has a mixed record. The first round of assessment done in 1991 resulted in a quantified score for each research project and the Center divested in few low-ranked projects, such as research on abiotic stresses and nematodes. The assessment done in 1995 confirmed what was already widely regarded as the most significant opportunities for impact, namely control of late blight disease in potato, improvement in seed quality and seed systems for both potato and sweet potato, and increasing dry matter yield and market opportunities in sweet potato. The results also gave direction for subsequent *ex post* impact assessment studies. In some cases, research that had been identified as high-pay-off investments in the 1995 *ex ante* assessment did not appear to be generating much adoption and became targets for studies on the constraints to adoption. Subsequently, results from these field studies downgraded the prospects for some technologies. Lower expectations for these technologies were reflected in scientists' assessments in the 2005 assessment round; prospects for true potato seed, for example, were markedly scaled back.

The additional emphasis given to modelling poverty impacts in the 2005 exercise suggested areas for programme realignment and where the Center should give priority for future *ex post* impact assessment to verify anticipated impacts. However, by the mid-2000s, CIP's capacity in impact assessment had seriously eroded. It is doubtful whether similar attention could be given to *ex post* assessment as a complementary tool for research planning unless this capacity is reinvigorated.

The quantitative assessments were less successful in helping the Center adjust its research portfolio during a period of sharply declining unrestricted funding for research. In the early 2000s, CIP was forced to downsize its staff and to seek restricted funding more aggressively (see Fig. 3.3). Staff reductions, mostly achieved through attrition, fell heavily on CIP's sweet potato, social sciences and regional positions. CIP maintained commitments to research on natural resource management and minor Andean root and tuber crops because of actual or perceived donor support for these projects, even though the priority assessments indicated these to be relatively low-pay-off areas. CIP's principal commodity programmes continued to weaken even as the Center established a new programme in urban agriculture. Rather than focusing research on 'high-pay-off' endeavours, the Center appears to have chosen to spread resources thinly across multiple endeavours in the hope of attracting new sources of financial support. Nevertheless, the publication of the 2005 priority assessment exercise (Fuglie 2007) was influential in the external review of the Center in 2007 and has probably contributed to a shift in new investment towards sweet potato.

Collion and Gregory (1993, p.29) point out that consensus building around programme priorities may be a more important outcome of an assessment exercise than the numbers themselves. This is most likely to be true in an environment of expanding and unrestricted research resources where strategic choices need to be made by the Center on where to expand programmes. However, in an environment of diminishing resources, it is more difficult to achieve consensus on programme priorities because programmes may be threatened. Scientists may become defensive about their own interests and find fault with the speculative nature of the exercise. Indeed, CIP's *ex ante* priority

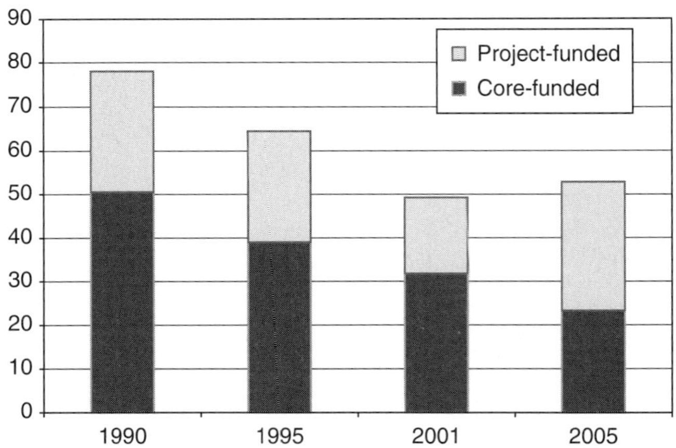

Fig. 3.3. The number of international research staff employed at CIP, 1990–2005.

assessments (like most *ex ante* assessments of research) likely exaggerate impacts. However, the value of the exercise is that it serves to highlight trade-offs among different research strategies by showing what research (and what assumptions about it) will give the highest relative impacts. Further, the exercise draws upon the collective judgement of the institute's scientific staff in doing so, rather than relying on the judgements of one or a small group of research managers.

A criticism of the approach to priority assessment outlined above has been its focus on economic welfare indicators and rates of return. While these remain powerful tools for identifying trade-offs among alternative research strategies, concepts like consumer surplus, NPV and rate of return may lack intuitive appeal among research managers and stakeholders. CIP's priority assessments sought to address this criticism by considering a broader range of impact indicators, including measures of poverty, health, employment and environmental effects of new technology (Table 3.6). But expanding the analysis to produce defendable measures such as these requires considerable investments in modelling and data and do not necessarily result in better decision tools Relying on qualitative scoring models to address other dimensions of poverty, while useful for raising awareness of, and attention to, these issues, is rarely sufficient to inform research managers of opportunity costs and trade-offs in allocating scarce resources to research.

In conclusion, it appears that the actual use of quantitative priority assessment of research in CIP's decisions has diminished over time. The increased reliance on restricted project funding has left less of the Center's programmes to the discretion of its management and staff. Nevertheless, in this changed context priority assessment retains a potentially important role in guiding the Center towards opportunities for restricted project funding that have the highest potential for impact on the poor. Additionally, CIP's experience shows that *ex ante* assessment works well in tandem with *ex post* impact assessment. The Center will need to reinvest in this capacity to ensure that research investment is oriented towards high-pay-off areas and to justify continued donor support for international agricultural research. CIP has taken some steps recently in this direction but a major limitation is the increasing dependence on the CGIAR as a whole on restricted funding with specific time-bound projects. It is intrinsically harder to fund either *ex ante* or *ex post* assessment as these need to take place outside of the time frame of any particular project. Finally, it seems to be more

Table 3.6. Anticipated impacts of CIP potato and sweet potato research by 2020. (From Fuglie, 2007.)

Adoption area of new technology ('000 ha)	3,753
Aggregate economic benefits (US$ million/year)	1,247
Benefits to rural poor (US$ million /year)	986
Number of person out of poverty ('000)	5,620
Rural employment generation ('000 jobs/year)	275
Human health impact (DALY saved/year)	21,048

difficult to use the priority assessment process to reach a consensus among Center staff in an environment where real resources for research are sharply declining. Nevertheless, even in this environment, the process can be an important tool for promoting understanding among Center staff and the broader stakeholder community when decisions need to be made to disinvest from particular projects.

References

Alston, J.M., Norton, G.W. and Pardey, P.G. (1995) *Science Under Scarcity: Principles and Practice for Agricultural Research Evaluation and Priority Setting*. Cornell University Press, Ithaca, New York.

Chilver, A., Walker, T.S., Khatana, V., Fano, H., Suherman, R. and Rizk, A. (2005) On-farm profitability and prospects for true potato seed (TPS). In: Razdan, M.K. and Matto, Matteo, A.K. (eds) *Genetic Improvement of Solanaceous Crops, Volume I: Potato*. Science Publishers, Enfield, New Hampshire, pp. 39–63.

CIP (1984) *Potatoes for the Developing World*. International Potato Center, Lima, Peru.

CIP (2004) *The CIP Vision: Preserving the Core, Stimulating Progress*. International Potato Center, Lima, Peru.

Collion, M.H. and Gregory, P. (1993) *Priority Setting at CIP: An Indicative Framework for Resource Allocation*. International Potato Center, Lima, Peru.

Fuglie, K.O. (2006) The impact of potato and sweet potato on poverty in Asia. In: Bourgeois, R., Svensson, L. and Burrows, M. (eds) *Farming a Way Out of Poverty: Forgotten Crops and Marginal Populations in Asia and the Pacific*. CAPSA Monograph No. 48, UN ESCAP Center for Alleviation of Poverty through Secondary Crops in Asia (CAPSA), Bogor, Indonesia, pp. 201–226.

Fuglie, K.O. (2007) *Research Priority Assessment for CIP's 2005–2015 Strategic Plan: Projecting Impacts on Poverty, Health, Employment and Environment*. International Potato Center (CIP), Lima, Peru.

Fuglie, K.O. and Yanggen, D. (2006) *Impact of Sweet Potato Biofortification on Vitamin A Deficiency in Developing Countries: An Ex Ante Economic Assessment*. International Potato Center, Lima, Peru.

Fuglie, K.O., Khatana, V., Ilangantileke, S., Singh, J.P., Kumar, D. and Scott, G. (2000) Economics of potato storage in Northern India. *Quarterly Journal of International Agriculture* 39(2), 131–148.

Herdt, R.W. and Hayami, Y. (1977) Market price effects of technological change on income distribution in semi-subsistence agriculture. *American Journal of Agricultural Economics* 59(2), 245–256.

Herrera, J. and Scott, G. (1992/93) Limiting factors for the production and use of potato: results of a survey of the national programs in Latin America. *Revista Latinamercana de la Papa*, 5/6(1), 122–131.

Horton, D. (1989) Constraints to sweet potato production and use. *Proceedings of the Workshop on Sweet potato Improvement in Asia, Trivandrum, India, 1989*. International Potato Center, Lima, Peru, pp. 219–223.

Koo, B., Pardey, P.G., and Wright, B.D. (2004) Saving seeds in the economics of conserving crop genetic resources. Ex situ in the Future Harvest Centers of the CGIAR, International Food Policy Research Institute, Washington, DC.

Stein, A.J., Meenakshi, J.V., Qaim, M., Nestel, P., Sachdev, H.P.S. and Bhutta, Z.A. (2005) Analyzing the health benefits of biofortified staple crops by means of the disability-adjusted life years approach: a handbook focusing on iron, zinc and vitamin A. *HarvestPlus Technical Monograph* Series 4, International Food Policy Research Institute, Washington, DC.

Walker, T.S. and Collion, M.H. (1997) *Priority Setting at CIP for the 1998–2000 Medium-term Plan*. International Potato Center, Lima, Peru.

Walker, T.S. and Crissman, C.C. (1996) Introduction and overview.Overview. In: Walker, T.S. and Crissman, C.C. (eds) *Case Studies of the Economic Impact of CIP-Related Technology*. International Potato Center, Lima, Peru, pp. 1–11.

Walker, T.S. and Fuglie, K.O. (2001) Impact assessment at the International Potato Center in the 1990s. In: *The Future of Impact Assessment in the CGIAR: Needs, Constraints, and Options, Proceedings of a Workshop Organized by the Standing Panel on Impact Assessment of the Technical Advisory Committee, 3–5 May 2000, FAO, Rome, Italy*. TAC Secretariat, Food and Agricultural Organization, Rome, Italy, pp. 37–41.

Walker, T.S. and Fuglie, K.O. (2006) Prospects for enhancing value of crops through public-sector research: lessons from experiences with roots and tubers. *Social Sciences Working Paper 2006–1*. International Potato Center, Lima, Peru.

4 The International Institute of Tropical Agriculture's (IITA) Experience in Priority Assessment of Agricultural Research

Victor M. Manyong, Diakalia Sanogo and Arega D. Alene

Abstract

This chapter documents International Institute of Tropical Agriculture's (IITA) experience in conducting priority assessment of agricultural research for development. For many years, IITA was defining its priorities mostly through consultation with its stakeholders. Following the recommendation of the Fifth External Programme and Management Review (EPMR) in 2001 to establish a sound methodology for priority setting, IITA developed two approaches. The first is the priority ranking exercise (PRE). It is an internal procedure using qualitative techniques, mainly for yearly prioritization of outputs and activities within a medium-term plan (MTP) project. The second is a formal priority assessment exercise (PAE). It is based on quantitative techniques, mainly to serve the Institute's long-term needs for priority assessment. The two approaches complement each other: so far, the formal PAE has been used to select and recommend strategic commodity programmes for the Institute while the priority ranking is used mainly for selecting better research operations within the priority commodity programmes. Strengths and limitations of each approach are discussed.

Keywords: IITA, priority ranking, priority setting, strategic commodity programmes

Introduction

Priority assessment is relevant for decision making and resource allocation at the institute level for investments across agro-ecological zones (AEZ) and projects, and at the project level where decisions about relative emphases across crops, breeding objectives and targeted environments apply. The International Institute of Tropical Agriculture (IITA) has a long tradition of assessing research priorities. For many years, IITA has used a bottom-up approach to define its research priorities across programmes and eco-regions. For example, the current IITA Strategic Plan for 2001–2010 is based on the strategic plans and priorities developed by the three subregional agricultural research organizations (SROs), including the Association for Strengthening Agricultural Research in Eastern and Central Africa (ASARECA), the West and Central African Council for Agricultural Research and Development (CORAF/WECARD) and the Southern Africa Center for Cooperation in Agricultural Research and Training (SACCAR). Each SRO prepared a strategic plan that involved an extensive priority assessment exercise (PAE) with broad consultation of many stakeholders. The SROs' strategic plans and priorities were further refined through analysis of the specific agricultural development challenges and research needs in the AEZ in which IITA is working to arrive at the current IITA Strategic Plan for 2001–2010 (IITA, 2001a,b).

The Fifth External Programme and Management Review (EPMR) of IITA appreciated the extensive effort that the Institute was making in its stakeholder (bottom-up) consultation process for priority assessment. In such a complex environment with many National Agricultural Research Systems (NARS) and other stakeholders, and a very fluid economic and political situation, a qualitative approach as used by IITA could yield realistic priorities. However, a quantitative, *ex ante* impact assessment priority evaluation process was needed to support the qualitative approach and provide the reasoning behind the stated priorities. Hence, one recommendation from the Fifth EPMR of IITA in 2001 was that the Institute should place immediate emphasis on establishing a sound quantitative methodology for prioritizing its research agenda. To implement this recommendation, IITA's Director General established a five-person Task Force on Priority Setting (TFPS). The specific terms of reference of the TFPS were (Manyong *et al*., 2001) the following:

1. Respond to the EPMR's recommendations on priority assessment at IITA.
2. Provide a more transparent and methodological approach to priority assessment.
3. Develop a set of criteria and procedures for evaluating project outputs and activities, in particular during the Institute's work planning week (WPW) in 2001.
4. Provide indicators to assist with the internal management of projects, especially to help project coordinators.

Formulation of a priority assessment approach

The Task Force reviewed approaches used by SROs and by some other Consultative Group on International Agricultural Research (CGIAR) Center (International Crops Research Institute for the Semi-Arid Tropics (ICRISAT), International Livestock Research Institute (ILRI) and International Maize and Wheat Improvement Center (CIMMYT)) in assessment of their priorities. It also studied the ISNAR recommendations for priority assessment on their web site. From this literature review, the TFPS identified 11 approaches that might be used in priority assessment. Each approach had its strengths and weaknesses. Based on this analysis, the TFPS recommended two approaches for assessment of priorities. One was a priority ranking exercise (PRE) and the second was a formal PAE.

The TFPS felt that a formal PAE would address the recommendation of the 2001 EPMR, but would require the use of many quantitative and qualitative data, most of which could be obtained only over longer periods than possible in the planning period. Experiences at other CGIAR Center (e.g. ICRISAT and ILRI) indicated that a formal PAE would require significant time and extensive consultations involving IITA's partners from NARS and other organizations, as well as additional human resources. The commitment of the Institute is also required for successful long-term priority assessment. The TFPS recommended priority assessment to become a new IITA priority time-bound output which could be produced under the Institute's Project C (assessing impact, formulating policy options and systems analysis).[1] Also, given the complexity of a formal PAE, the 2003 Center Commissioned External Review (CCER) for Social Science at IITA recommended that the PAE be conducted initially for Nigeria because of the relative importance of this country in sub-Saharan Africa. Results from the PAE in Nigeria could then serve as a model for other countries, subregions or the entire continent.

With respect to internal planning within the Institute, the panel recommended a PRE. A PRE combines peer review and scoring. Basically, the exercise requires the use of simple qualitative indicators that are then subjected to quantitative analysis. It can be conducted quickly. The PAE and PRE are complementary in that a PAE leads to the identification of priority commodity programmes while a PRE assists in defining operations within the prioritized commodity programmes. The remainder of this chapter explains the process and examples of the PRE conducted at IITA, and the results of the formal PAE conducted in Nigeria. The last section discusses the strengths and weaknesses of a PRE and a PAE, drawing on the IITA experience.

PRE: Process and Findings

The purpose of the PRE was to serve the internal needs of IITA's projects for priority assessment among various activities and outputs. It was designed as a management tool to serve scientists and research managers and to guide prioritization within a medium-term plan (MTP) project.

[1] Since 2001, priority setting has been a time-bound activity in all the IITA projects that address social science issues, although the name of the project has changed over time. Currently, priority setting is an output target embedded in the MTP project referred to as opportunities and threats.

The PRE process

The TFPS used a participatory approach to achieve the following two objectives during the PRE:

1. Define criteria suitable for priority ranking of outputs and activities.
2. Provide the basis for the calculation of weights to be attached to the identified criteria.

The TFPS developed a semi-structured questionnaire, using suggestions from IITA management about the list of potential criteria for priority assessment and results from informal discussions with scientists from each research division. The survey form contained the following two questions:

Question 1: Please list and give weight to criteria you consider important in setting priorities among outputs and research activities to contribute to IITA's mission.

Question 2: Describe in one paragraph the procedure (i.e. set of steps) you wish IITA would implement in setting priorities among outputs and research activities.

The aim in the first question was to develop a list of criteria for a PRE and to suggest the weight for each suggested criterion. At IITA, research operations are undertaken within research projects. Each research project is subdivided into research outputs and each output is made up of research activities. Therefore, two sets of criteria for PRE needed to be developed: one for outputs and one for research activities. In the second question, the aim was to receive suggestions on how a PRE can best be implemented at IITA.

The questionnaire was mailed to scientists at the IITA stations in Cotonou, Benin Republic; Ibadan, Kano and Onne in Nigeria; Namulonge in Uganda; and Yaoundé in Cameroon. An example of the questionnaire is included in Appendix 4.1.

Criteria and weights for priority ranking (Question 1)

Data generation and analytical techniques

In total, 35 scientists returned the survey forms and suggested 38 criteria. Some of them were very similar. Therefore, the criteria were pooled into a final set of 18 (Appendix 4.2). Some criteria are related to IITA's mission with respect to food security, poverty reduction, sustainable management of agroecosystems and capacity building (relevance to the organizational mission). The second set reflects requirements to smoothly implement research operations at IITA (feasibility). The third set corresponds to expectations of stakeholders (impact). As one can see, the suggested criteria are nested. For example, criteria on food security and poverty reduction or meeting farmers' needs are not mutually exclusive. To address this problem and select a few criteria out of the 18, the TFPS used three methods to rank the criteria and compared results from the methods.

Method 1 considered the frequency of responses for each criterion. Criteria with a high frequency of response (regardless of the score) were selected; i.e. criteria that were recommended by a high number of scientists.

Method 2 used the mean score weighted by the frequency of positive responses (i.e. mean weight from scientists' scores); i.e. the product of mean score and frequency of responses with scores more than 0 over the total number of responses. This method combined the relative importance scientists gave to a criterion as shown by the mean score, and the popularity of that criterion from the proportion of scientists who chose it.

This combination is shown in the following equation where s_{ij} is the score given by respondent i ($i = 1, \ldots, 35$) with respect to criterion j ($j = 1, \ldots, 18$), and N is the total number of responses. The mean score Xj weighted by frequency for criterion j is calculated as:

$$W\overline{X_j} = \frac{n_j}{N} * \overline{X_j} = \frac{n_j}{N} * \frac{\sum s_{ij}}{N}, \tag{4.1}$$

where n_j represents the number of respondents with a weight > 0 for criterion j.

Method 3 consisted of analysing the data using principal components analysis. The sum of loading scores for the major components (which explain more than 50% of variation) was used to rank criteria. The three methods were applied first to the suggested criteria for research outputs, then to those criteria suggested for research activities.

Criteria and weights for research outputs

Each of the three methods resulted in a similar ranking for the first-ranked six criteria, with a minor difference in their ordering (see Appendix 4.3). In addition, these six criteria had the highest mean scores. Therefore, the TFPS recommended adopting them as criteria for the priority ranking of research outputs and their mean scores as their weights.

Criteria and weights for research activities

In selecting criteria at the activity level, those chosen as criteria for outputs were not used. Methods 1 and 2 were deemed most appropriate for ranking criteria at the activity level. Of the remaining 11 criteria, the TFPS recommended the first seven, and their mean scores were used in the PRE at the activity level. The cut-off point for choosing a criterion was that at least 50% of the respondents had chosen it. Results on the highest-ranked criteria for research activities are shown in Appendix 4.4.

Recommended steps for PRE: synthesis of responses (Question 2)

The following steps were recommended for the smooth implementation of PRE.

1. Research outputs:
   - Define researchable issues, challenges and opportunities at subregional and/or national level and discuss with stakeholders.
   - Identify key constraints, challenges and opportunities with respect to the project purposes and goals and group them under common themes.

- Identify outputs for which IITA has a comparative advantage.
- Set a clear time frame for each research output.
- Identify and address gaps in partnerships.
- Prioritize and rank research outputs within each project.

2. Research activities:
   - Devise researchable activities that contribute significantly to research outputs for which IITA has a comparative advantage.
   - Submit research protocols for each activity. The research protocol contains details related to criteria for the PRE, allowing experts to proceed with priority ranking.
   - Prioritize activities on the basis of research protocols within each output using the agreed criteria.

Implementation of PRE

To implement the PRE, a form was developed and sent to scientists involved in a project in the form of an Excel spreadsheet to facilitate keying of data and analysis. In the columns of the Excel sheet, there are criteria and assigned weight (the sum of weights for all criteria equal to 100), and in the rows, there are research activities. Each scientist assesses each research activity on a scale of 1–10 (1 being low and 10 being high) for its contribution to each of the criteria indicated in columns. Using formulas in the Excel spreadsheet on the basis of Eq. 4.1, a total score is calculated for each research output from all the forms completed by scientists. Each overall score is translated into a percentage by dividing that total score by the total maximum achievable score (i.e. the total scores if each scientist had scored 10 in each cell of the form). The research output with the highest percentage is ranked first, and so on for others. In addition to the overall ranking of research outputs, the results allow to compare scores of all the research outputs on a single criterion column-wise and to assess the highest and least contribution of a given research output to various criteria row-wise. The results from PRE are useful to set priorities but also to identify strengths and weaknesses of research outputs, which could lead the research managers/team to design appropriate strategies to correct for weaknesses of their research projects. The same approach applies for priority ranking of the research activities using milestones for the exercise.

Utilization of PRE approach at IITA

At the beginning of the effort in 2001, leaders of various IITA projects were enthusiastic to apply the PRE approach to prioritize their research outputs during the IITA annual planning meetings. Application of the approach was made possible by the gathering of many scientists in one location during the IITA annual planning meetings and because each project had only between three and seven research outputs. While the duration of the annual meetings was adequate to rank research outputs for each project, it became obvious that there was not enough time to implement the same exercise for research activities. The second reason for not implementing PRE for research activities was that scientists often had to go back to their stations and revisit their research activities after prioritizing their project research outputs. Therefore, PRE was never implemented for research activities.

In the past two years (2007 and 2008), the PRE approach has received renewed interest from both IITA scientists and management because IITA had to reorganize its research portfolio into a new framework under MTP projects. Under the new MTP format, each project is made up of strategic objectives (SOs) and each SO is made up of tactical objectives (TOs). One or more research activities can be conducted under each TO. Some modifications were introduced to make PRE easier and quicker. First, only five outcome criteria that more or less describe the IITA mission were considered for PRE. In Appendix 4.3, criteria 1, 2, 3 and 5 were retained. A new criterion was added on the contribution to enhance the quality of science, since it is a major expected outcome from an international agricultural research institute such as IITA. Criteria 4 and 6 in Appendix 4.3 were dropped. Second, a modification was made on the quantification of the weight to be attached to each criterion, which is now determined through a participatory approach by scientists. Finally, PRE was applied once for both TOs and SOs. The results were first calculated for TOs. The aggregated data of all the TOs were pooled into one score for each SO. Scores for each SO were used to rank them. Therefore, the final output from this new PRE was a ranking of TOs for each SO and ranking of SOs at the level of the MTP project.

The approach has been used successfully as a formal and non-biased system for priority ranking for five out of the six MTP projects of IITA: roots and tubers systems, cereals and legumes systems, agro-biodiversity, agriculture and health and opportunities and threats. An example of such an exercise can be found in Manyong *et al.* (2009). Under the new PRE, prioritization of research activities for each TO remains the domain of individual scientists.

Lessons from the PRE

The PRE has resulted in creating greater awareness among all scientists of the critical importance of priority assessment to ensure that scarce research resources are allocated in ways that will have the greatest impact on the poor. It has reinforced IITA's commitment to long-term priority assessment. A computer-based system was developed to assist scientists and research managers in the implementation of the PRE.

PRE is participatory, easy and quick, compared to a formal PAE. However, it is still largely based on 'expert' opinion and it applies primarily qualitative techniques. Also, PRE seems to be best for prioritizing research operations within a MTP project. There remains the need to develop an appropriate approach for prioritizing research activities.

PAE: the Case of Nigeria

An example of applying a more formal, quantitative priority assessment approach is described in the recently published work by Alene *et al.* (2007a), which was used to assess agro-ecological strategic research priorities based on potential impacts of alternative commodity research programmes to reduce poverty in Nigeria. The same approach can also be found in Alene *et al.* (2007b). This approach aims at ranking commodity research programmes to guide strategic investments by IITA. Unlike PRE, this PAE analysis was conducted at a level of aggregation higher than research outputs. While PRE was developed and applied to rank research operations within an already prioritized commodity research programme, the quantitative PAE of commodities in Nigeria was used to prioritize and recommend commodity research programmes that best contribute to poverty reduction. Therefore, the two approaches are complementary: PAE results in priorities for commodity research programmes, while PRE defines priorities for operations within each of the selected commodity research programmes.

Methodology

The methodology for this priority assessment approach is described in detail in Alene *et al.* (2007a) and is summarized in Appendix 4.5. A model of income determination for small-scale agricultural producers was used to assess income changes associated with the adoption of agricultural technologies and the resulting yield changes in each AEZ. Income was defined to be the sum of farm income, off-farm income and monetary and in-kind transfers to a household. Two types of data were needed for the household-level poverty analysis: household survey data and forecasted changes in yields, production costs and adoption of technologies. Household survey data collected in 2001 and 2002 from a nationally representative sample of 3180 households were used to estimate the household-level impacts of agricultural research. The sample households were drawn using multistage stratified random sampling of 12 Nigerian states which are representative of the three major AEZs: dry savannah, moist savannah and humid forest. Data were collected on a range of variables including household characteristics, agricultural production and input use, off-farm incomes, monetary and in-kind transfers, food and non-food expenditures and adoption of modern inputs.

A total of 144 scientists (i.e. research and extension workers) and research managers from IITA and more than ten NARS were involved in generating the research and technology data through extensive discussions and interviews using a detailed set of questions. Expert opinions were elicited regarding likely changes in important parameters, such as yields and production costs, based on the detailed information provided on a set of technological interventions. Key research programme areas of the respective NARS were identified and stock was taken within each programme of the production and postharvest technological innovations that might be developed if sufficient research budgets were available. The household income data from a nationally representative sample of 3180 households and data on projected yield and production cost changes from experts' opinion were used to derive the various poverty measures. The study considered four scenarios based on specific net changes in crop yields and expected technology use intensity. Two sub-approaches were applied: an efficiency-based approach calculating estimated economic benefits using economic surplus analysis and an equity-based approach.

Summary of key findings

The results of the efficiency-based model indicate that most research programmes have internal rates of return (IRR) greater than 50%. The commodity research programmes were categorized into high-, medium- and low-priority groups. The top commodity programme was yam with a total net present value (NPV) of US$7.5 billion and an IRR of 131%. The high-priority programmes in the order of decreasing importance were yam, cassava, maize, rice, cowpea, citrus fruits, sorghum, plantain, poultry, millet and groundnut. Generally, leafy vegetables, fruits and livestock were medium-priority programmes, whereas other vegetables and industrial crops were low-priority programmes. Of course, there were variations within commodity groups. For instance, citrus among fruits and poultry among livestock were high-priority commodities. It should be noted that the categorization of commodity programmes into different priority groups is subjective, depending on how many programmes the available research resources can support and hence, which programmes would be of high priority and should be funded and which ones would be of low priority and receive little or no support. If, for instance, resources allow only one commodity to be supported, the second priority programme would then become low priority.

The aggregate impact of increased agricultural research on poverty reduction depends on the magnitude of the shift in rural income distribution associated with the entire research portfolio. Figure 4.1 presents the cumulative distribution of rural incomes with and without increased agricultural research for each AEZ. The distribution shows an outward shift in incomes with increased agricultural research, with noticeable differences across AEZ. The income shift is more significant in the moist savannah, moderate in the humid forest and marginal in the dry savannah. It is interesting to note that the income shift is marginal in all zones for the 10% of low-income households, indicating that the income changes following increased research would not be sufficient to lift the poorest of the poor out of poverty.

From the equity-based approach, maize research has the largest potential impact on poverty reduction with poverty incidence reduced by 5% following increased maize research. Therefore, maize is the highest-priority commodity programme. The high-priority commodity programmes in order of decreasing importance are maize, cowpea, rice, yam, cassava, millet, sorghum, groundnut and poultry. Generally, most livestock commodity programmes, fruits, vegetables and industrial crops have little or no impact on poverty and are of low priority.

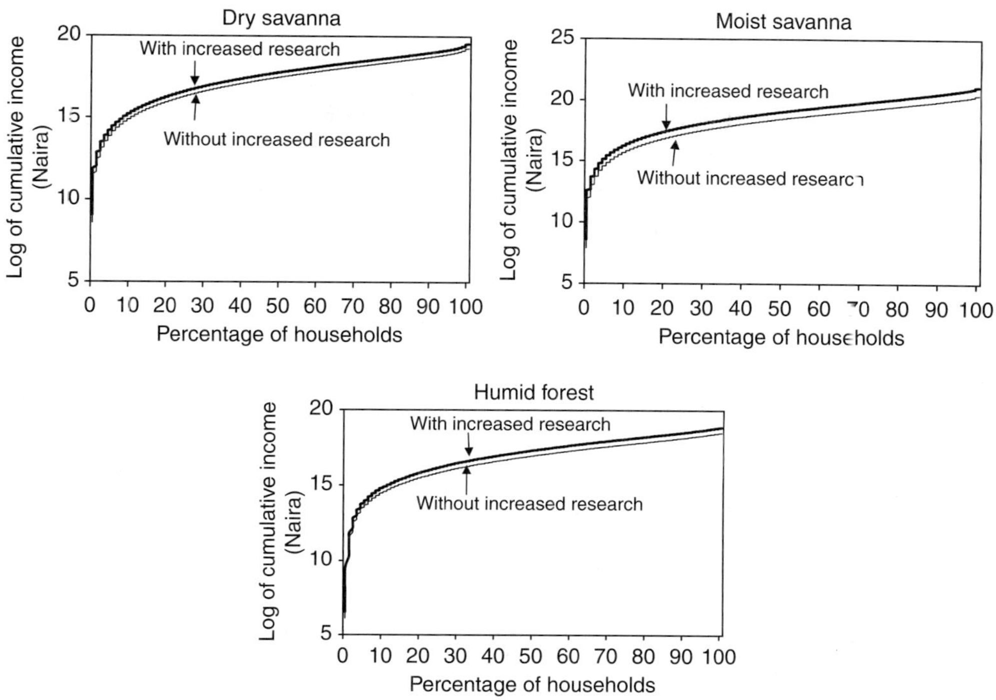

Fig. 4.1. Rural income distribution by AEZ in Nigeria with and without increased agricultural research.

Potential impacts of most research programmes on income inequality are negligible. While most poverty-reducing commodity programmes also have inequality-reducing impacts, some can actually increase inequality. Maize research has the highest inequality-reducing impact, indicating that it is grown by most of the poor including the poorest of the poor. As mentioned earlier, the Gini coefficient for rural Nigeria before research was 0.572 and this falls to 0.554 following the forecasted income changes from increased maize research. Cowpea, millet, sorghum, groundnut and soybean also slightly reduce inequality. On the other hand, cassava, yam and cocoa research programmes slightly increase inequality. Cocoa research has a negligible effect on poverty but increases income inequality, indicating that much of the producer benefits from this research programme will be captured by the better-off farmers.

It should be noted that the national poverty line for Nigeria of US$0.75 per capita per day was applied for this assessment, as established and used by World Bank (1996) and FOS (2004).

The issue of efficiency–equity trade-offs was addressed using rank correlation and graphical analyses of research priorities identified on the basis of efficiency and equity criteria. First, the correlations of the priorities according to the efficiency and the equity criteria were examined using the Spearman's rank correlation coefficient (ρ). The results indicated a similar set of five top commodities that would simultaneously satisfy the objectives of efficiency and poverty.

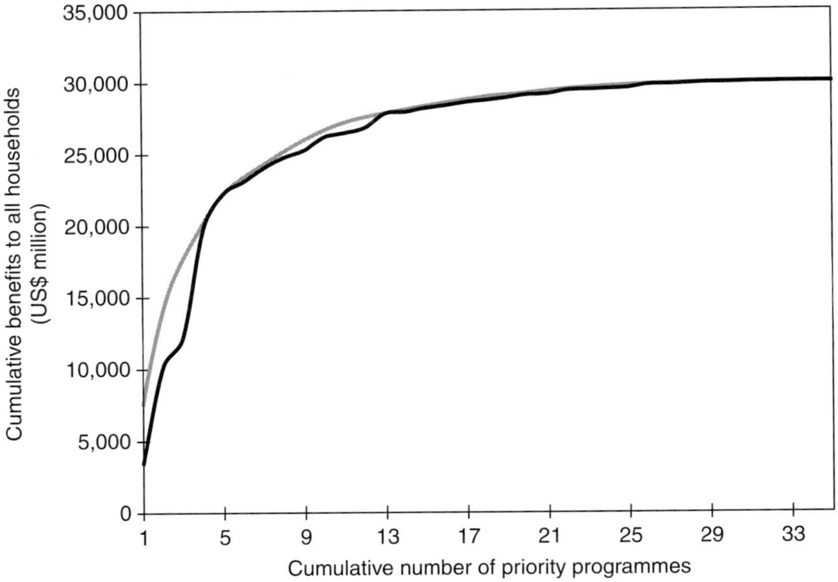

Fig. 4.2. Cumulative benefits to all households from adding research programmes according to efficiency and equity criteria.

A proper analysis of the magnitude of the efficiency–equity trade-offs should go beyond rank correlations. Figure 4.2 illustrates the costs to society in terms of forgone benefits, measured as the shortfall of benefits to all households following the allocation of research resources in ways that maximize benefits to the poor as opposed to maximizing benefits to all households. It is shown that societal benefits from pursuing efficiency and equity objectives converge and this convergence is more or less sustained after the top five commodity programmes have been included in the research portfolio. The overall effect of prioritizing on the basis of equity on societal benefits is only 3%, relative to the societal benefits from efficiency-based priorities. That is, society would forgo only 3% of total research benefits if agricultural research programmes were prioritized on the basis of equity. However, analysis of forgone benefits to society alone does not reveal all that is of interest regarding the efficiency–equity trade-offs. It is also important to know the marginal benefits to the poor from prioritizing research programmes according to the equity criterion (i.e. benefits to the poor). Figure 4.3 illustrates how the total benefits to the poor from the equity-based research portfolio compare with the efficiency-based portfolio. The benefits to the poor from the equity-based research portfolio are consistently higher than the efficiency-based portfolio. The analysis indicates that the poor would capture 24% of total research benefits, compared to their share of total production of 29%. The overall effect of prioritizing on the basis of equity on research benefits to the poor is 8%. That is, the poor would capture 8% more research benefits from equity-based research resource allocations than they would from efficiency-based allocations. The relative importance of the marginal benefits to

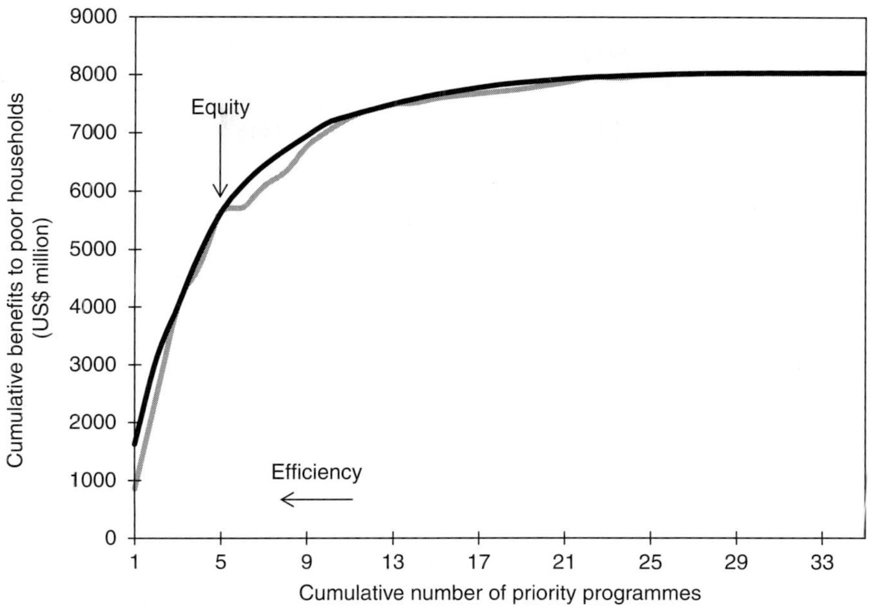

Fig. 4.3. Cumulative benefits to poor households from adding research programmes according to efficiency and equity criteria.

the poor from equity-based priorities and the lack of significant efficiency–equity trade-offs indicates possibilities to direct research benefits to the poor without compromising research efficiency.

Lessons from priority assessment of research commodities in Nigeria

Nigeria is a large country in sub-Saharan Africa (SSA) and implementing PAE proved to be a useful experience. The exercise showed that it is possible to conduct a formal PAE for major programmes/commodities. The major challenge is the availability and accuracy of information required to set priorities. Often, the responsiveness of experts in supplying technical information was less than optimal. A formal PAE requires substantial resources (financial, time, human) for it to be successfully implemented.

Limitations and Constraints of PRE and PAE

PRE

A key weakness of the PRE is that it is based on the opinions of experts who do the scoring of research operations. An inherent limitation of such an expert opinion-based research evaluation is the subjectivity involved in estimating the likely effects of commodity-specific research. Researchers may be optimistic about the techniques they are developing, and overstate yield increases while understating cost increases. Second, the PRE sets priorities among SOs that pertain to one MTP project. Different experts who belong to different projects assess their research outputs independently. Some of the experts modified weights to adapt to the specifics of their MTP projects. While the changes might be appropriate, it makes it difficult to compare outcomes from PRE applications across various MTP projects. Third, PRE was not effective at the level of research activities. A more user-friendly tool is needed for this level of detailed priority assessment.

PAE

Formal quantitative priority assessment requires substantial resources (human, finances, methods, time, etc.) to be successfully completed. The process involves quantitative studies that need to be extended at lower levels to set priorities among MTP projects, SOs and TOs within MTP projects and research activities within TOs. This could be a daunting exercise. The scientists' bias in estimating benefits and costs of new technologies already mentioned for PRE also applies to PAE. The utility of findings from such studies depends on the institutional commitment of research managers.

While the scientific value of the exercise has been recognized through publication of its findings in refereed journals, it is not certain how researchers, managers (national and international institutes), funding agencies and other stakeholders will use the recommendations of this formal PAE. Nevertheless, as long as continued pressure on public research budgets requires research to concentrate on fewer projects and programmes with high returns and potential for poverty reduction, sustained efforts need to be undertaken to develop appropriate tools and approaches for priority assessment to effectively rank all research projects and programmes.

Linkages between a quantitative PAE and a qualitative PRE

Our experience with conducting PAE and PRE is that PAE should logically inform the implementation of PRE. In other words, priorities need to be set first for the research commodities through a more rigorous quantitative PAE periodically and PRE is conducted yearly within prioritized research commodities to set priorities among research operations. As it was written earlier in this chapter, PAE is very costly compared to PRE. Therefore, PAE could be conducted with the aim of establishing a long-term strategic plan (e.g. 10 years for IITA), while PRE would be conducted yearly to fine-tune research operations. However, this should not be considered as a one-way exercise, because experiences from yearly PREs would be useful as well in designing new PAEs.

References

Alene, A.D., Manyong, V.M., Tollens, E.F. and Abele, S. (2007a) Targeting agricultural research based on potential impacts on poverty reduction: strategic programme priorities by agroecological zone in Nigeria. *Food Policy*, 32, 394–412.

Alene, A.D., Manyong, V.M., Tollens, E.F. and Abele, S. (2007b) Setting agricultural research priorities based on potential impacts on poverty reduction: the case of Nigeria. Pages 8–26. In: Alene, A.D., Manyong, V.M., Abele, S. and Sanogo, D. (eds) *Assessing the Impact of Agricultural Research on Rural Livelihoods: Achievements, Gaps, and Options*. Scientific papers presented during the twenty-sixth conference of the International Association of Agricultural economists, 12–18 August 2006, Brisbane, Australia. IITA, Ibadan, Nigeria, 80.

Federal Office of Statistics, Federal Republic of Nigeria (FOS) (2004) *Nigeria Living Standards Survey, 2003/04*. Federal Office of Statistics, Abuja, Nigeria.

International Institute of Tropical Agriculture (IITA) (2001a) *Delivering the African Agricultural Development Potential: The IITA Vision. Strategic Plan 2001–2010*. IITA, Ibadan, Nigeria, 20.

International Institute of Tropical Agriculture (IITA) (2001b) *IITA Strategic Plan 2001–2010: Supporting Document*. IITA, Ibadan, Nigeria, 41.

Manyong, V.M., Diels, J., Dixon, A., Hughes, J. and Kormawa, P. (2001) *Priority Setting at IITA: Report by the Task Force on Priority Setting*. IITA, Ibadan, Nigeria, 30 (mimeo).

Manyong, V.M., Okechukwu, R., Ojiambo, P. and Asiedu, R. (2009) Setting priorities: case study of IITA's root and tuber crops systems programme. *African Journal of Roots and Tuber Crops* (in press).

World Bank (1996) *Nigeria – Poverty in the Midst of Plenty: The Challenge of Growth with Inclusion*. A Poverty Assessment Report, World Bank, Washington, DC.

Appendix 4.1

Survey form for scientists

Dear colleague,

I believe you have seen from a DG's e-mail that we are in the process of re-arranging the IITA projects, outputs and activities (going from 14 to 6 projects). In response to the EPMR's criticism that priority setting in IITA lacks some transparency, RPEC (= the highest IITA research management body at the time of the survey) has now established a small task force to come up with a set of criteria and a procedure to prioritize our research. The short-term aim is to agree on those criteria and procedures we can use at WPW to prioritize outputs and activities in the new 6 projects. This task force would like to have input from all colleagues. We will appreciate your spending a few minutes of your time to urgently respond to only two questions below. Send your reply (e-mail or drop a hard copy) to Victor Manyong, Chairman, Task Force on Priority Setting, Building 401, IITA Ibadan (E-mail: v.manyong@cgiar.org). Thank you.

Stakeholders' Participation in the Development of Criteria and Procedures for Setting of Research Priorities at IITA: The Contribution from Scientists

Scientist Name: ....

The IITA's mission is to enhance the food security, income, and well-being of resource-poor people primarily in the humid and subhumid zones of sub-Saharan Africa, by conducting research and related activities to increase agricultural production, improve food systems, and sustainably manage natural resources, in partnership with national and international stakeholders.

Question 1: Please list and give weight to criteria you consider as important in setting priorities among outputs and research activities to contribute to the IITA's mission (use suggestions provided).

Question 2: Describe in one paragraph the procedure (i.e. set of steps) you wish IITA implement in setting priorities among outputs and research activities.

Output
Step 1 ...
Step 2 ...
Step 3 ...
...
Step *n* ...

Research activities
Step 1 ...
Step 2 ...
Step 3 ...
...
Step *n* ...

Appendix 4.2

Potential criteria to be used for priority ranking of outputs and research activities at IITA

Potential criteria	Weight for outputs (0–100)	Weight for activities (0–100)
1. Contribution to enhancing food security		
2. Contribution to poverty alleviation and improving human well-being		
3. Contribution to sustainable resource management		
4. Contribution to global environmental benefits and to alleviating adverse ecological conditions, particularly climate change		
5. Contribution to alleviating the negative impacts of HIV/AIDS and other endemic diseases		
6. Contribution to the generation of International Public Goods		

Continued

Potential criteria	Weight for outputs (0–100)	Weight for activities (0–100)
7. Amount of resources available (financial/human) to implement task(s)		
8. Balance between cost of, and benefits from, research		
9. Probability of success		
10. Time frame (duration) required to achieve expected result(s)		
11. Internal logic among research activities (linkage between tasks)		
12. Comparative advantage of researchers (expertise) in implementing task(s)		
13. Comparative advantage of the future technology among alternative policy conditions and ecologies		
14. Contribution to multidisciplinary team building		
15. Contribution to capacity building and strengthening of NARS, and implementing effective partnerships		
16. Contribution to meet farmers (needs) interest and other users needs		
17. Contribution to meet regions interest: Solving a problem(s) of a regional priority as identified by subregional organizations (SRO)		
18. Contribution to meet donors' interest		

Appendix 4.3

Ranking and weight of criteria using three alternative methods for the output level

		Method*				
No.	Criterion	1	2	3	Mean score	Weight (%)
1	The contribution to enhancing food security	27	11.0	4.09	13.5	22
2	The contribution to poverty alleviation and improving human well-being	30	13.4	8.16	14.7	24
3	The contribution to sustainable resource management	29	7.3	2.29	8.3	13
4	The contribution to meet farmers' and other users' needs	29	10.8	9.67	12.3	20
5	The contribution to capacity building and strengthening of NARS and implementing effective partnership	25	4.7	2.14	6.2	10
6	The probability of success	22	4.3	2.91	6.5	11
Total					61.5	100

*Method 1: Frequency of responses; Method 2: Mean score weighted by proportion of positive responses; Method 3: Principal component analysis.

Appendix 4.4

Selected criteria and their weight for priority ranking of research activities

No.	Criteria	Method 1*	Method 2*	Mean score	Weight (%)
1	Researchers have expertise and comparative advantage to implement activity (technical expertise, critical mass, experience, training opportunities/development of technical expertise to achieve activity)	22	6	4.2	20
2	Favourable balance between cost and benefits from research (*Ex ante* analysis/impact assessment, development costs, value of outputs justify level of inputs, efficient use of resources)	21	5.1	3.4	16
3	Meets donors' interest (donor focus country, donor focus interest (within IITA's agenda), appropriate partnerships, donor/stakeholder appreciation of the problem-solving activity)	20	4.3	2.7	13
4	Generates international public good(s) (publications/technologies/methodologies, shared benefits (many countries/many people), freely available and useful to the general public, 'common good of humanity')	19	4.2	2.5	12
5	Contributes to solving problem(s) of a regional priority as identified by SROs (Regional significance, addresses SRO priorities, regional network linkages)	19	4.0	2.4	12
6	Researchers have the resources (financial and human) or can get the resources to implement the activity (Facilities/equipment, availability of funds, sufficient resources among stakeholders, attractive to donors)	18	5.7	3.2	15
7	Activity will be achieved in a relatively short time (relative to alternative methods that can be used to reach same results) (realistic milestones/defined stages, verifiable indicators, activity achievable within lifetime of output)	18	4.3	2.4	12
Total				20.8	100

*Methods 1 and 2 (see Appendix 4.3); Cutting-off point for choice of criteria: at least 50% of respondents have chosen the criterion.

Appendix 4.5

Quantitative priority setting for Nigeria: detailed analytical method (From Alene *et al.*, 2007.)

Despite the growing concern about, and interest in, explicitly considering poverty alleviation in research planning, conceptual and methodological challenges have limited priority setting based on the potential impacts of agricultural research on poverty alleviation. Most priority setting works have used the economic surplus approach (Alston *et al.*, 1995) and emphasized efficiency objectives by prioritizing research programmes according to the net present value (NPV) of projected research benefits (e.g. Mills, 1997; Nagy and Quddus, 1998; Mutangadura and Norton, 1999). In particular, the challenge relates to the complexity of the links between agricultural research and poverty alleviation (Kerr and Kolavalli, 1999). Kerr and Kolavalli (1999) review the available evidence and conclude that the poor are simultaneously producers, wage-earners and consumers, with technological change having complex direct and indirect and often offsetting effects on their real incomes. In SSA, where a significant share of farm output is consumed on-farm, recent empirical work has shown that the direct impacts on producers – in the form of increased home consumption – are more important than the indirect effects on consumers (de Janvry and Sadoulet, 2002).

With household-level data, income growth associated with crop-specific yield changes can be aggregated to create measures of changes in poverty. A model of income determination for small-scale agricultural producers facilitates the determination of such income changes associated with the adoption of agricultural technologies and the resulting yield changes in each AEZ (Alwang and Siegel, 2003). Income is defined to be the sum of farm income (I), off-farm income and monetary and in-kind transfers to a household. For the *i*th household in the *j*th AEZ (suppressing all subscripts for notational simplicity), farm income can be defined as:

$$I = H(YP–C), \qquad (1)$$

where H is a vector of hectares of land allocated to each of the crops or number of heads of livestock raised, Y a diagonal matrix of yields, P a vector of prices and C is a vector of per hectare or per animal costs of production. Changes in farm incomes can be decomposed as:

$$\Delta I = \Delta H(YP–C) + H\Delta YP – H\Delta C \qquad (2)$$

Equation 2 shows the four major effects that commodity-specific research has on household income and poverty. The first is through changing the allocation of land to each crop or the number of heads of each livestock type raised. Research produces new technologies that change the relative profitability of crops and livestock and induce land reallocations or livestock mixes. The second is the change in yields due to the new technology. The third component of the effect of research is the effect of changed supplies on prices received by farmers. Price changes depend on the tradability of the commodity, which is reflected in the elasticity of demand. With more tradable commodities, research-related supply shifts are not likely to affect producer prices (Alston *et al.*, 1995). Moreover, increasing population, such as that in Nigeria, could counteract the negative price effects of new technologies (Smale and Heisey, 1994). The final effect in Eq. 2 is through the impact of research and the new technologies that generally require more inputs and hence greater costs per hectare or per animal than traditional technologies. Exceptions in this regard include technologies such as Integrated Pest Management (IPM) that reduce chemical applications and hence reduce costs per hectare. Alwang and Siegel (2003) associate the final effect of research with per hectare cost decreases. It should be noted, however, that while research reduces per unit production costs, it generally increases per hectare costs.

From the four major effects of agricultural research on farm incomes, the effects through yields, prices and cost changes are the most important. In the economic surplus framework, the yield and cost changes are used to produce the supply shift. On the other hand, changes in land allocations and livestock compositions are difficult to predict and can only be examined in an *ex post* fashion (Alwang and Siegel, 2003).

The expected changes in yields, prices and costs can be used to compute the expected research-induced changes in farm incomes for the *i*th household as follows:

$$E(\Delta I_i) = H\Delta YP + HY\Delta P - H\Delta C, \text{ if adopter of new technology}$$
$$=0, \text{ if non-adopter of new technology} \qquad (3)$$

In this chapter, the computation of the expected income changes was guided by the following steps and basic assumptions. First, it was assumed that there would be no price changes (i.e. $\Delta P = 0$) due to increasing market liberalization and growing domestic food demand in Nigeria. Second, researchers' estimates were used to predict commodity-specific yield (ΔY) and per hectare production cost changes (ΔC). Third, since farmers must adopt the research product for research benefits to be realized, the assumption was made that their current adoption of purchased inputs (fertilizer, herbicides, pesticides and improved planting materials) implied that technologies to be developed by the research system would also be adopted. Household-specific information on adoption of purchased inputs was obtained from a nationally representative sample survey of rural households, which revealed that 41% in the dry savannah, 65% in the moist savannah and 53% in the humid forest adopted modern inputs. Finally, income changes associated with increased commodity research were then computed for each household using Eq. 3. The expected changes in farm incomes were then added to initial household incomes and the resulting incomes were compared to a poverty line and aggregated to form expected changes in poverty incidence. Following Foster *et al.* (1984), poverty measures were derived as:

$$\frac{1}{N}\sum_{i=1}^{q}\left(\frac{z - y_i}{z}\right)^0, \qquad (4)$$

where N is the total number of households, q is the number of poor households, y_i is the income or expenditure of the ith poor household and z is the poverty line. It is important to note that technology adoption and the changes in net yields and incomes should not be expected to occur at one time. Rather, it would be approximately 25–30 years (Collinson and Tollens, 1994; Alston *et al.*, 1995) after technology generation before technology adoption reaches maximum levels and research benefits are realized fully.

References (for Appendix 4.5)

Alene, A.D., Manyong, V.M., Tollens, E. and Abele, S. (2007) Targeting agricultural research based on impacts on poverty reduction: strategic program priorities by agro-ecological zone in Nigeria. *Food Policy* 32(3), 394–412.

Alston, J.M., Norton, G.W. and Pardey, P.G. (1995) *Science Under Scarcity: Principles and Practice for Agricultural Evaluation and Priority Setting*. Cornell University Press, New York.

Alwang, J. and Siegel, P.B. (2003) Measuring the impacts of agricultural research on poverty reduction. *Agricultural Economics* 29, 1–14.

Collinson, M. and Tollens, E. (1994) The impact of the international agricultural center: quantification and interpretation. *Experimental Agriculture* 30, 395–419.

de Janvry, A. and Sadoulet, E. (2002) World poverty and the role of agricultural technology: direct and indirect effects. *Journal of Development Studies* 38(4), 1–26.

Foster, J., Greer, E.J. and Thorbecke, E. (1984) A class of decomposable poverty measures. *Econometrica* 52, 761–776.

Kerr, J. and Kolavalli, S. (1999) Impact of agricultural research on poverty alleviation: conceptual framework with illustrations from the literature. In: Proceedings of the ETPD Discussion Paper 56. International Food Policy Research Institute, Washington, DC.

Mills, B.F. (1997) *Ex ante* agricultural research evaluation with site specific technology generation: the case of sorghum in Kenya. *Agricultural Economics* 16, 125–138.

Mutangadura, G. and Norton, G.W. (1999) Agricultural research priority setting under multiple objectives: an example from Zimbabwe. *Agricultural Economics* 20(3), 277–292.

Nagy, J.G. and Quddus, M.A. (1998) National agricultural commodity research priorities for Pakistan. *Agricultural Economics* 19, 327–340.

Smale, M. and Heisey, P.W. (1994) Maize research in Malawi revisited: an emerging success tory? *Journal of International Development* 6, 689–706.

5 Priority Assessment at the Center for International Forestry Research (CIFOR): Confronting the Challenges of a Policy-oriented Natural Resources Management Research Portfolio

David A. Raitzer

Abstract
This chapter reviews the history of systematic attempts to select research priorities for the Center for International Forestry Research (CIFOR), beginning with the four sets of consultative processes used to identify the initial priorities suggested for the Center at establishment, of which CIFOR's first (1996) strategy selected a subset. In 2000, CIFOR attempted to undertake a systematic priority assessment approach through application of a derivative of the Research Application Area Framework developed by Australia's Commonwealth Scientific and Industrial Research Organization (CSIRO). Although the approach helped to clarify trade-offs among research topics and regions, it had limited success in fostering a more clearly focused research agenda. Between 2000 and 2007, the Center relied on consultative and implicit priority assessment methods, as it developed programme strategies and regional strategies. In this context, scientists reported that they chose priorities primarily on the basis of knowledge gaps and perceived problem prevalence. During 2007, CIFOR made another attempt at more systematic priority assessment, so as to inform a new strategy for 2008–2018. This consisted of a five-step qualitative exercise, which involved defining a long list of research alternatives, elaborating on the details of research themes and impact pathways, applying a Delphi process to discuss and rank alternatives, using a structured scoring process to compare alternatives regarding specific criteria and finally selecting priority topics.

Keywords: Priority setting, priority assessment, NRM research, policy research, impact pathways

Introduction

There has long been recognition of the potential benefits of systematic priority assessment for informing the development of CIFOR's research agenda. At the same time, the Center's portfolio is dominated by policy-oriented research on forest-related issues. Policy-oriented research poses challenges for the prediction of adoption potential, while forestry and natural resources management (NRM) research compounds problems of benefit quantification and valuation. In combination, these challenges have impeded the development of systematic methods for appraising the comparative impact potential of research alternatives.

Despite these constraints, there have been a number of attempts to integrate priority assessment into the selection and design of CIFOR's research themes. These approaches have been principally qualitative and have utilized differing arrays of implicit and explicit selection criteria.

Overview of Methods Applied to Date

Early priority identification exercises

Before the Center was established, an extensive series of exercises was conducted to assess the merit of adding an institute with a forestry mandate to the Consultative Group on International Agricultural Research (CGIAR), as well as to appraise the relevance of alternative research topics and modalities to the system's goals. These efforts were all essentially priority assessment exercises, although appraisal against impact-related criteria varied in transparency among the processes.

The origins of these exercises can be traced to the 1988 International Task Force on Forestry Research (ITFFR), sponsored by the Rockefeller Foundation. This assessment (ITFFR, 1988) is presented as a review of 'key areas of research which have contributed in the past, and could contribute in the future, to achieving objectives for each of the four Tropical Forestry Action Plan (TFAP) areas of opportunity' (sustainable land use and food security; energy security, sustainable forest industry development and tropical forest ecosystem conservation and management). Actual analysis of past 'contributions' did not refer to documented research benefits. Rather, recent research results which illustrated problem prevalence were qualitatively reviewed and summarized to infer these potential 'contributions'. Five key priorities were identified for a forthcoming international forestry research 'council': agroforestry/watershed management; natural forest ecology and management; tree breeding and tree improvement; utilization and marketing; and policy and socio-economics. About 130 topics were suggested across these five major research areas, and relative priority was not assigned among these.

The CGIAR Technical Advisory Committee's (TAC) Forestry Panel Chair and CGIAR Secretariat subsequently undertook a consultative exercise to identify more specific priorities within each of the five areas (Gregersen and Spears,

1989). In selecting these areas, emphasis was stated to be given to the following criteria: '(i) what the area of research has contributed and could contribute in the future, (ii) how it could benefit from international centralization; and (iii) what complementary activity/investment is needed'. The analysis was unique in that it attempted to summarize benefits documented *ex post* for each research area, so as to identify past and potential contributions. This process resulted in the identification of the following topics for potential research 'centralization': biotechnology, vegetative propagation, biological nitrogen fixation, germplasm conservation, policy research, utilization research, forest entomology/pathology and remote sensing.

The CGIAR donors decided to conduct their own parallel exercise for defining priorities through consultations with the CGIAR 'Forestry Working Group'. The group identified four perspectives on deforestation and land use, to which four categories of research activities were related: environment and forest conservation, production forestry, social forestry and agroforestry (CGIAR Forestry Working Group, 1991).

In 1991, the Australian Center for International Agricultural Research (ACIAR) was appointed by the CGIAR to implement the establishment of CIFOR, and also took on the mantle of identifying potential research priorities for the future institute (ACIAR, 1992). ACIAR's team convened 22 national and regional seminars and surveyed a wide variety of individual scientists and research institutes to elicit topics of research interest. Background papers were commissioned from teams of experts for 14 topics. The draft CIFOR strategy prepared by ACIAR combined insights from these papers with the findings of the ITFFR, TAC Forestry Panel and the CGIAR's Forestry Working Group, as well as additional consultations by the United Nations' Food and Agriculture Organization (FAO) and the World Bank. Criteria for the selection of specific priorities were not explicitly specified. This strategy proposed scores of topics in four areas: management and rehabilitation of uplands for stability; tropical forest management for sustainability and conservation; germplasm conservation, genetic improvement and plantation management; and utilization and marketing of forest products.

The early consensus-building approach employed for CIFOR resulted in a longer list of topics than the Center could feasibly address with available resources (Anderson *et al.*, 1995). This conforms to Nilsson *et al.*'s (2004) observations that 'using consensus building to identify research priorities can result in overly broad and vague guidelines'. In the absence of a clear ranking of alternatives, the consensus solution was all inclusive.

Assessment of priorities for the 1996 strategy

In 1995, CIFOR attempted to define options for more systematic priority assessment, so as to support the development of the Center's first strategy. A number of potential criteria were discussed for the screening of research options, and there was consideration of establishing impact-related criteria for proposal approval. Ultimately, the development of systematic priority assessment was postponed. CIFOR's first strategy hence declares (CIFOR, 1996):

> To ensure that CIFOR research will make a difference, high priority is given to the development of a rigorous system for the planning and management of research activity. CIFOR's research evaluation system aims to link priority setting and resource allocation to the assessment of impact using careful monitoring of research activities.

The ten priority projects identified in the 1996 strategy were a selection of the longer list of researchable issues identified by ACIAR. A matrix of 20 assessment criteria is presented in the strategy document, but application of this is not detailed. The strategy's annex of priorities follows a logical framework approach, in which problems to be addressed through research, as well as impact pathways, are qualitatively described. However, it is unclear whether this was the basis of project selection.

In the years that followed, CIFOR developed a template approach for documenting intended 'impact pathways', which was encapsulated in 'project strategies' (Spilsbury, 2000a). Such templates defined intended outputs, users and beneficiaries, but were not used for systematic comparison of research options. To complement these pathways, the Center attempted to draw attention to audience characteristics and information needs through 'market research' approaches. Thus, these studies attempted to identify forestry research products of other institutes that had fostered extensive impact. General principles about how research findings can be effectively targeted and disseminated were also derived and discussed in a number of papers and presentations (Spilsbury and Kaimowitz, 2000; Spilsbury, 2001).

Attempt to Apply the Research Application Area Framework

In 2000, CIFOR attempted to undertake a more systematic priority assessment approach through application of a derivative of the Research Application Area Framework initially developed by the US Industrial Research Institute, and adapted by Australia's CSIRO (Healy, 1999). The approach involves generating semi-quantitative estimates of the 'impact potential' and 'feasibility' of research in a particular area. These two estimates are displayed graphically to provide guidance to the decision maker. Four sub-criteria for the two main criteria were defined, as per below.

Impact potential:

- potential benefits from products and services:
    - maximum additional benefits (economic, environmental, social, capacity enhancement) from successful research.
- implementation context for products and services:
    - likelihood of the results of successful research being used to achieve desired outcomes.

Feasibility:

- scientific potential:
    - scope for growth in knowledge in the relevant scientific fields and improvements in research tools and techniques.

- research capacity:
  - CIFOR's ability to assemble internationally competitive research teams to deliver research outputs.

Assessment of these criteria was based on an eco-regional approach. As background, a data book was compiled on trends in resource use and importance for eight regions. This review included data on poverty, demographic trends, extent and environmental status of forest resources in the region and the social importance of forest resources. Descriptions were also developed for five eco-regional zones (termed 'areas of research opportunity'). These included trends, descriptions of forest benefits, a strengths-weaknesses-opportunities-threats analysis, a collection of potential themes and a description of capacity for each zone.

A 'regional needs assessment' was conducted by scoring poverty and nutritional needs, environmental states and livelihood dependency on forests for each region. Eco-zones and programmes were subsequently qualitatively scored against each of the four criteria for impact potential via Delphi approaches. Fifty-one participants, including nine members of management, 23 research staff, 11 board members and eight observers, participated in the scoring workshop.

The method was successful in identifying trade-offs among areas of research opportunity. Figures 5.1 and 5.2 depict some of the key trade-offs explored. The workshop also ultimately resulted in recommended changes to resource allocation among areas of research opportunity and regions. One particularly novel result was the accordance of high priority to the 'lowland moist long dry' area within Central and West Africa, even though this was an area in which CIFOR had no activity at the time (CIFOR, 2000).

However, the results of the process were never fully adopted, as Center management stated concerns that the approach was overly mechanical and rigid. A possible explanation for part of this perception is the unit of analysis. An eco-regional approach is relevant for adoption domains that are bounded by eco-regions. However, policy adoption contexts (governments and institutions) are not necessarily well defined by eco-regional zones. As noted by Spilsbury (2000b):

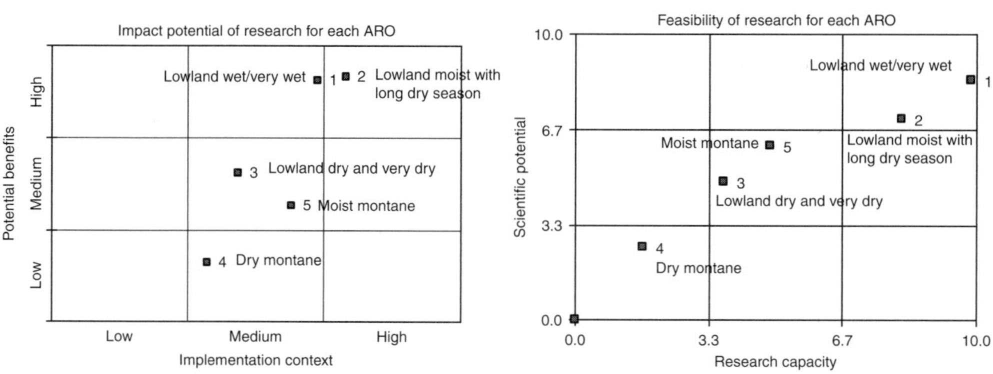

Figs. 5.1. and 5.2. Graphical descriptions of trade-offs among priority assessment criteria resulting from CIFOR's use of the Research Application Area Framework. (From Spilsbury, 2000a.)

> The issue of selecting appropriate problem sets (AROs) upon which to set priorities was a key issue. Participants had difficulty in using ecolimatic zones as a means discriminating between problem sets. Many strategic research issues are common across these zones.

This meant that each eco-region contained a diverse range of research topics, and these topics overlapped among eco-regions. Audi and Mills (1998) note that 'ambiguity about the set of [research] alternatives to be compared' can cause problems for appraisal against priority assessment criteria. In CIFOR's policy-oriented context, it is possible that the eco-regional basis of comparison introduced such ambiguity.

Consultative approaches

From 2000 to 2007, CIFOR relied on consultative, qualitative and implicit priority assessment criteria, as it has redefined programme strategies and developed strategies for each region. These criteria were embodied as scientists developed research plans in conjunction with programme directors, prospective donors and partners.

In 2003, senior scientists at CIFOR were interviewed regarding the considerations that drive selection of research priorities at the individual level (Raitzer, 2005). These interviews were conducted in a semi-structured format, in which respondents were queried about future research priorities, after which the basis for the selection of the priorities was explored.

The selection of research emphases was primarily driven by a combination of intuition, creativity, observations and problem perceptions according to responses received. Perceived problem prevalence was tied with interest/intuition as the most commonly cited consideration during the selection of envisioned future research emphases, with 40% of respondents indicating that each had an important role (n = 21; Fig. 5.3). The next most commonly claimed considerations were the needs that partners had articulated for information and perceived gaps in existing knowledge (both with 35% of respondents). Following this was the need for continuation of past research, as prior results necessitated follow-up and/or further action so as to result in useful outputs (with 20% of respondents). Finally, funding availability was cited as an important factor by 10% of respondents.

Similarly, consultative deliberations of perceived problem prevalence, knowledge gaps and comparative advantage were used to develop the 'strategies' for CIFOR's three programmes during 2003 and subsequent years (Raitzer, 2005).

While the considerations enumerated by CIFOR's scientists and management included important aspects of impact potential, it is also clear that many factors important to understanding whether research is likely to effectively contribute to the Center's mission were not among the factors reported. For example, the probability of research success in addressing the knowledge gap, the likelihood of large-scale influence/adoption and the magnitude and nature of potential benefits were not reported. According to the interviews, adoption

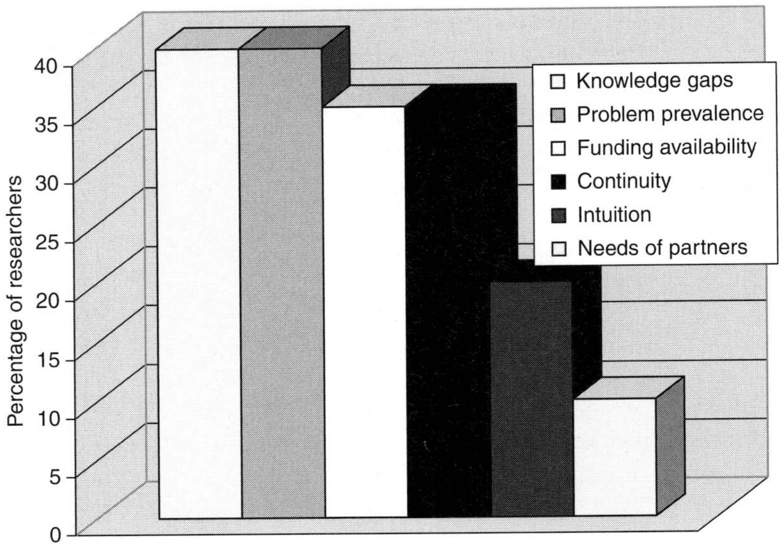

Fig. 5.3. Information reported to be considered during the development of individual future research interests and priorities at the Center for International Forestry Research (*n* = 21). (From Raitzer, 2005.)

potential appears to be more explicitly considered in the context of selecting means for the dissemination and communication of research results, rather than as a determinant of research themes (Raitzer, 2005).

Selection of Research Priorities for CIFOR's 2008–2018 Strategy

During the development of CIFOR's most recent strategy, there was recognition of the need for more systematic selection of priorities, which was reinforced by a recommendation of CIFOR's 2006 External Programme and Management Review to clarify priority-setting procedures. Accordingly, the Center worked towards developing a multi-criteria method for appraising a long list of potential research alternatives. The process of priority assessment followed three principal stages.

Preparatory stage – defining the research alternatives and assessment criteria

First, a preliminary 'long list' of research topics – or domains – was developed by CIFOR scientists on the basis of three sets of inputs: (i) a revised articulation of CIFOR's mission and goals; (ii) an analysis of the external environment in which CIFOR operates; and (iii) suggestions on potential topics elicited from stakeholders and partners through direct interviews and an online survey. Thirteen candidate 'domains' were identified by groups of scientists within the Center, who considered the above inputs, along with their own scientific interests in an inductive process.

The next preparatory step was to develop a 'template' for further defining and elaborating a set of 'candidate' research domains (Box 5.1). Internal consultations were used to define elements of the domains to develop, after which teams of scientists worked to develop 'domain descriptions', as per this narrative template. The draft domain narratives were shared among staff and were individually discussed and refined during specific sessions on each domain during CIFOR's Annual Meeting in August 2007, which attempted to further elucidate details on intended outputs, users, outcomes, impacts and beneficiaries.

Due to time limitations, quantitative analysis of prospective impacts was not possible. Moreover, such analysis would have been difficult due to the fact that research details were primarily defined at the domain level, which involves several millions of US dollars per year of annual investment. This aggregate level often groups together rather diverse specific research topics and intended outputs, and in CIFOR's policy-oriented context, each specific output has many potential user groups and impact pathways. This means that estimation of quantitative measures of impact at such an aggregate level requires analysis of a prohibitively large array of diverse possible events. In lieu of this, the intention was to focus on more qualitative discussion of the feasibility, magnitude and importance of likely impacts of alternative options, as a prelude to scoring.

Box 5.1. Template for strategic research area (domain)

1. Context
2. Challenges and opportunities
3. Overall goal or strategic result
4. Scientific value, expected outcome(s) and pathways to impact
5. Rationale for CIFOR to work in this research domain
6. Demand and target clients/users
7. Anticipated impacts:
   - equity/empowerment;
   - contribution to well-being of the poor;
   - expected economic benefits;
   - extent of environmental benefits (not including any monetary benefits assessed above).
8. Geographic focus
9. Time frame
10. Feasibility:
    - fundability;
    - organizational capacity;
    - assumptions and risks;
    - overall probability of success.

Research themes under domain:
- descriptive title;
- research problem;
- key questions and hypotheses;
- intended outcomes.

Scoring overall preferences via a Delphi approach

The second stage was to use a Delphi approach to elicit the rationales for internal perceptions and preferences among scientific staff regarding the long list of key research domains. The original intent of the Delphi process was to score the domains against a set of assessment criteria related to feasibility and impact potential, which were identified through consultation with CIFOR staff, and to explore the rationales for specific scores. However, due to time constraints, and the orientation of the hired consultant moderator, this approach was altered to indicate overall scores of domains.

The Delphi was moderated externally and carried out throughout September 2007. It comprised an iterative, anonymous process by three separate panels. Two of these were composed of CIFOR scientists (one including only globally recruited scientists with international expertise; and a second consisting of more locally specialized staff), while a final panel included senior management and two Board members. All 13 domain narratives and a set of reference selection criteria were provided to panellists, who were asked to score the overall degree to which specific domains merit inclusion into CIFOR's future research portfolio. Three iterations took place during which panellists scored, provided the rationales for scores and reviewed the rationales and scores from other panellists.

The content of rationales put forward under the process for domain preferences only partially accorded with the list of assessment criteria provided for reference. Most of the arguments articulated related to 'feasibility' criteria, such as organizational capacity, topical relevance to forests, fungibility and fundability. A substantial share of statements also related to whether a topic was broad enough, or was an appropriate organizational unit, for research. Only a small proportion of stated rationales for scores directly related to criteria listed under impact potential, such as quantities or qualities of benefits likely to be created, or whether adoption or use of the research was likely.

Five research domains consistently were scored highly with high intra-panel agreement, while five others were consistently given low scores. Three domains were subject to a wide divergence of opinion, which did not move towards consensus over successive iterations for the panel of globally recruited scientists with more international exposure (Table 5.1).

Scoring alternatives against assessment criteria

The third stage consisted of a structured scoring exercise by the Strategy Steering Committee members against criteria identified by CIFOR scientists and management. These included the scale of potential benefits, relevance to the three areas of CIFOR's revised mission (contribution to well-being of the poor, equity and environmental protection), fundability and complementarity/spillover benefits to the rest of CIFOR's work (Table 5.2).

Seven strategy steering committee members submitted scores, out of the 11-member group. So as to include the full spectrum of scores in the overall

Table 5.1. Final mean scoring (where 1 is most important to include and 6 is least important) and level of agreement within panels across all panels engaged. Panel 1 includes CIFOR Global Scientists, while Panel 2 includes more location specific CIFOR scientific staff, and Panel 3 includes CIFOR Board members and management.

Domain	Domain title	Panel 1 agreement	Panel 2 agreement	Panel 3 agreement
1	Maintain and expand forest-based carbon sinks	Disagreement (4)	Strong agree-ment (3)	Strong agreement (1)
2	Achieve livelihood improvements and forest conservation in response to climate change (adaptation)	Weak Disagreement (3)	Weak agreement (1)	Strong agreement (1)
3	Guide the biofuels revolution for forest conservation and development	Weak agreement (5)	Weak agreement (3)	Weak agreement (3)
4	Ensure sustainable income generation from small-scale and community forestry	Strong agreement (1)	Strong agreement (1)	Weak agreement (2)
5	Harness industrial-scale forest plantations for sustainable development	Disagreement (4)	Weak agreement (4)	Weak agreement (4)
6	Achieve conservation and development in integrated landscape management	Weak agreement (2)	Weak agreement (2)	Strong agreement (2)
7	Put forests to work for water resources management	Weak agreement (5)	Strong agree-ment (5)	Strong agreement (6)
8	Address corruption and weak governance through the forestry sector	Weak Disagreement (4)	Weak agreement (5)	Strong agreement (6)
9	Build capacity to deal with contemporary resource management issues in national forestry systems	Strong Disagreement (3)	Strong agreement (5)	Strong agreement (6)
10	Recognize, protect and improve the contributions of forests to human health	Weak agreement (5)	A spread (4)	Strong agreement (5)
11	Trade and investment	Weak agreement (2)	Weak agreement (2)	Weak agreement (2)
12	Manage degraded and deforested land	Strong Agree-ment (5)	Weak agreement (5)	Weak agreement (5)
13	Sustainable management of tropical production forests	Weak agreement (3)	Weak agreement (3)	Weak agreement (3)

Table 5.2. Ranking of research alternatives by mean normalized scores for assessment criteria received from Strategy Steering Committee Members. The composite score reflects equal weighting of all criteria.

Rank	Composite		Poverty		Environment		Equity		Funding		Spillovers to CIFOR	
1	11. Trade and investment	0.91	11. Trade and investment	1.00	1. Maintain and expand carbon sinks	1.00	4. Sustainable income from small-scale and community forestry	1.00	1. Maintain and expand carbon sinks	1.00	11. Trade and investment	1.00
2	1. Maintain and expand carbon sinks	0.82	4. Sustainable income from small-scale and community forestry	0.88	13. Sustainable management of tropical production forests	0.83	11. Trade and investment	0.98	2. Climate change adaptation	0.97	1. Maintain and expand carbon sinks	0.87
3	4. Sustainable income from small-scale and community forestry	0.80	13. Sustainable management of tropical production forests	0.70	11. Trade and investment	0.82	2. Climate change adaptation	0.87	3. Guide the biofuels revolution	0.94	2. Climate change adaptation	0.87
4	2. Climate change adaptation	0.79	2. Climate change adaptation	0.67	6. Conservation and development in landscape management	0.73	6. Conservation and development in landscape management	0.80	8. Corruption and weak governance in forestry	0.74	4. Sustainable income from small-scale and community forestry	0.84
5	3. Guide the biofuels revolution	0.75	6. Conservation and development in landscape management	0.63	3. Guide the biofuels revolution	0.70	3. Guide the biofuels revolution	0.75	11. Trade and investment	0.73	6. Conservation and development in landscape management	0.84
6	6. Conservation and development in landscape management	0.73	5. Forest plantations for sustainable development	0.61	5. Forest plantations for sustainable development	0.58	13. Sustainable management of tropical production forests	0.75	4. Sustainable income from small-scale and community forestry	0.70	9. Build capacity in national forestry systems	0.81

7	13. Sustainable management of tropical production forests	0.73	3. Guide the biofuels revolution	0.60	2. Climate change adaptation	0.57	10. Contributions of forests to human health	0.68	6. Conservation and development in landscape management	0.67	13. Sustainable management of tropical production forests	0.81
8	5. Forest plantations for sustainable development	0.62	10. Contributions of forests to human health	0.60	4. Sustainable income from small-scale and community forestry	0.56	1. Maintain and expand carbon sinks	0.67	5. Forest plantations for sustainable development	0.64	3. Guide the biofuels revolution	0.77
9	8. Corruption and weak governance in forestry	0.62	1. Maintain and expand carbon sinks	0.56	12. Manage degraded and deforested land	0.53	8. Corruption and weak governance in forestry	0.61	10. Contributions of forests to human health	0.64	8. Corruption and weak governance in forestry	0.75
10	10. Contributions of forests to human health	0.58	8. Corruption and weak governance in forestry	0.55	9. Build capacity in national forestry systems	0.47	5. Forest plantations for sustainable development	0.58	13. Sustainable management of tropical production forests	0.58	5. Forest plantations for sustainable development	0.71
11	9. Build capacity in national forestry systems	0.54	9. Build capacity in national forestry systems	0.45	7. Water resources management	0.45	12. Manage degraded and deforested land	0.56	7. Water resources management	0.52	12. Manage degraded and deforested land	0.65
12	12. Manage degraded and deforested land	0.53	12. Manage degraded and deforested land	0.41	8. Corruption and weak governance in forestry	0.43	9. Build capacity in national forestry systems	0.51	9. Build capacity in national forestry systems	0.48	10. Contributions of forests to human health	0.58
13	7. Water resources management	0.47	7. Water resources management	0.39	10. Contributions of forests to human health	0.41	7. Water resources management	0.50	12. Manage degraded and deforested land	0.48	7. Water resources management	0.52

aggregate measure, mean scores were used for data analysis.[1] In general, the criteria measured distinct aspects of the research domains, as the correlation between average scores was low, with the exceptions of 'equity/empowerment' and 'poverty/material well-being', which had a correlation of 0.90, and 'overall scale of benefits' and 'spillover benefits to other CIFOR research', which had a correlation coefficient of 0.85.

Total impact expected at a given time could be considered to be a product of the contribution per area and scale (area affected) of benefits. For this reason, the scale and contribution per area scores for the three categories of benefits were merged into combined measures, so as to compare overall total contributions to different areas of CIFOR's mission. Thus, the scores for total contributions to poverty, environment and equity in the areas of influence were weighted by scores for the overall scale of likely benefits. These scores were then normalized on a 0–1 scale. In addition, a composite ranking was calculated, based on equal weighting of the five measures (20% for overall contributions to poverty, environment and equity, as well as fundability, and benefits to other CIFOR research). In general, the rank of composite scores for domains matched the outcomes of the Delphi process.

Given the restricted time available for the scoring process following the Delphi approach, there was limited opportunity for providing the rationales for scores or for debating the underpinning assumptions. This was a weakness of the approach, as assumptions regarding research use were not explored, and the potential for priority assessment to fulfil an internal learning function was not fully fulfilled.

More generally, while the process helped the Center to make important progress towards systematic and transparent selection of priorities, it did not fully explore potential opportunities to embed external input from partners and outside experts. This meant, in some cases, that research options outside of CIFOR's present core capacity were less fully explored than those within. It also meant that important opportunities to use the priority assessment process as a means to embed partner perspectives and improve consensus about CIFOR's most effective role were not fully realized.

Initial operational effects

The Strategy Steering Committee, taking into account the results of the second two stages, selected six research domains for inclusion in the strategy. These six were among the top seven composite scored domains and all received average scores that were in the top half of those received in the Delphi exercise. Thus, the selected domains generally accorded with the priority assessment results.

As of late 2008, it is too early to determine the effects of the latest exercise on the specific activities within CIFOR's portfolio. The domains generally

[1] Medians and means were calculated for the scores, but there was little difference between the two.

represent an evolution, rather than a revolution, of CIFOR's prior research portfolio, and all domains that were prioritized generally reflect the prior interest of multiple Center scientists. Moreover, as each domain is broadly bounded, there remains room for the topics excluded from the shortlist of selected domains to potentially re-emerge as activities within the priority domains.

Follow up definition of impact pathways

At CIFOR's 2008 Annual Meeting, participatory exercises were piloted with scientists to reinforce the Center impact culture and to help embed consideration of impact potential in the prioritization of activities within domains. These 90 min exercises were a form of impact pathway analysis, which began with a review of the characteristics of policy-oriented research that had substantial documented impact. These included the case studies that reported quantitative estimates of benefits attributable to policy-oriented research in CGIAR Science Council (2008) and in Raitzer and Ryan (2008). The fact that all the cases involved policy analysis research (rather than more upstream science) included early, and in most cases extensive, engagement with intended users or target policy makers and explicitly (or implicitly in one case) quantified the benefits expected from suggested policy changes was presented to the scientists.

The presentation was followed by a three-step exercise. In the first step, scientists were asked to identify a single important target policy/management decision for each of CIFOR's 16 MTP 'outputs'. The policy target was to have an envisioned policy change as a result of research, which should be implemented, such that there is bearing on environmental, social or economic goals.

The second step was the participatory construction of network maps of actors influencing the target policies, in a somewhat similar manner to the social network maps employed in Participatory Impact Pathways Analysis (PIPA, described in Chapter Two of this volume). These maps included the target policy in the Center, with actors as boxes, relationships among the actors as arrows and the likely receptiveness of different actors to evidence from CIFOR indicated for each actor. In contrast to PIPA, however, these maps did not focus on the establishment of new relationships, but on the use of existing relationships for information and influence flows.

The third step consisted of narrative description of the research 'logic'. 'Five As' were enumerated for each output (the term was coined in one of the first exercises by CIFOR Associate Arild Angelsen). These included:

- **A**im: problem/opportunity and target policy/management reform to rectify the situation;
- **A**ctors: those influencing the target policy;
- **A**llies: the subset of actors who would be likely to be receptive to the expected research findings;
- **A**mmunition: evidence from CIFOR research that could be used by allies to advocate for the target policy/management change;
- Plan of **a**ttack: a narrative description linking the previous four As.

In general, scientists' reaction to the exercise was positive, and it was possible to observe stated target policies evolve through discussion of these elements, along with the intended 'evidence'. A number of scientists attested the fact that the probability of influence had not been previously considered in as much detail. In this sense, the exercise was perhaps more successful in fostering reconsideration of assumptions underpinning likely impacts than was the more structured priority assessment exercise for the strategy. However, the impact pathways exercise was illustrative, rather than comprehensive, as it only covered a single policy target per more than US$1 million of annual research investment. To truly shift the portfolio, more comprehensive assessment is planned, with evaluation intended to follow-up on the veracity of projected assumptions.

Key Future Challenges

CIFOR has continually attempted to move towards systematic assessment of research options, so as to focus on topics with high-impact potential. Yet, in so doing, it has encountered a number of methodological challenges that have constrained these efforts.

First, there is the challenge of identifying a proper unit of analysis for internationally relevant policy-oriented research. If the unit of analysis is too small, the list of research alternatives becomes cumbersome, and the ultimate detailed list of priorities can restrict scientific creativity. If the unit of analysis is too large, somewhat disparate activities may be amalgamated, and it can be difficult to identify concrete research products and associated intended impacts. Moreover, research alternatives with broad and amorphous bounds are more likely to overlap, rendering comparison more difficult.

Elicitation of the probability of research success and adoption is an issue that has received scant attention in the priority assessment literature (Anderson, 1992). However, this element is crucial to the reliability and utility of a priority assessment process. Priority assessment is conventionally assumed to base estimates on the expert judgement of scientists involved in the potential research alternatives. There are obvious conflicts of interest involved, which are likely to bias predicted adoption upwards. This raises the need to challenge bias and unrealistic assumptions. However, the effectiveness of alternative means and necessary conditions to effectively challenge and validate assumptions have not been sufficiently analysed in the literature to date.

Provided that unbiased parameter estimates can be elicited, the context of policy-oriented research presents challenges for identifying the impact potential of research alternatives. Even in the comparably simpler context of technology-oriented research, successful science does not necessarily translate into successful impact on goals at the level of poverty alleviation or environmental protection. Policy-oriented research compounds these difficulties, as there are usually a greater number of intermediate adoption events than for technology-oriented research (such as crop genetic improvement; SPIA, 2006). For example, a successful case of policy-oriented research impact will often first depend upon advocacy by others. In turn, decision makers may alter assistance programmes or change rules governing NRM practices. In response to the reform, resource management practices are shifted, so that productivity increases and/or environmental degradation is reduced. As a result, the quantity, sustainability or distribution of benefits is changed. Thus, accurately predicting the probability of impact requires prediction of the probabilities of a protracted chain of events – a task that many would rather avoid due to the heroic assumptions involved.

At the same time, effective application of policy- and management-oriented research findings often requires substantial local implementation capacity, as findings are intended to be applied in local regulatory frameworks and the actions of other agencies. Consequently, if regulations are not enforced due to limited capacity, ineffective institutions and/or problems of corruption, the impact of such research will be constrained. This suggests that more, rather than less, attention should be paid to the probability of substantive influence for policy-oriented research. A systematic priority assessment exercise can offer important possibilities to increase understanding of how these constraints affect the adoption potential of alternative research themes, but only if underlying assumptions are made explicit, and are subject to discussion.

Forest policy and management research also poses problems due to the exceptionally long time frames concerned. Research, adoption and benefit lags are all exacerbated by the long gap between changes in practice and consequences for the production of products and services. The further into the future impacts are projected, the more uncertain the context for those impacts becomes. If it is accepted that successful influence is a realistic assumption, additional challenges remain. Many of the intended benefits from CIFOR's research, such as enhanced environmental services or equity, are not always amenable to valuation with conventional economic surplus techniques. Defining systematic processes to appraise relative contributions to these goals requires innovation and new approaches.

At the same time, the fact that CIFOR receives public funds designated for development creates an obligation to ensure and demonstrate that research activities are likely to contribute to development goals. As it currently stands, there is uncertainty about the level of benefits generated by international natural resources and policy-oriented research to date (Kelley and Gregersen, 2003; Lele *et al.*, 2003; Maredia and Raitzer, 2006). This places a certain 'burden of proof' on the expected impacts from continued or expanded investment in the areas of research upon which CIFOR focuses. Investment in explicit priority assessment is one important way to mitigate this uncertainty.

Given these challenges, the pursuit of systematic priority assessment requires considerable planning and organizational commitment. There is no fast-and-easy solution to predict the impact potential of many policy-oriented research options, but research choices can benefit from more validation of whether intended outcomes are likely and exploration of their relative importance and magnitude. Substantial time and effort are needed to design appropriate mechanisms to expose assumptions to external evidence and insight and to investigate a wide array of potential impact pathways. Appreciation of these time requirements and careful planning may be essential to ensure the substantive success of priority assessment as a learning tool for a research organization.

Conclusion

CIFOR's priority assessment has to date relied on qualitative and consultative processes to appraise potential areas of research. Early efforts were highly consultative, applied varying sets of criteria and ultimately resulted in a more extensive list of priorities than the Center could feasibly address. A subsequent attempt in 2000 to use ecological zones for the basis of priority assessment was perceived as 'mechanical'. Recently, the Center devoted a greater degree of concerted effort in developing an appropriate process for developing and ranking priorities, and has made important progress towards systematic assessment of research alternatives. A recent impact pathways exercise helped to foster some reconsideration of factors affecting policy influence among Center scientists. Yet, there remains some room for improvement in future exercises, so as to generate more novel information to enlighten scientists and embed enough external input to foster additional reflection over key assumptions.

References

ACIAR (1992) *Center for International Forestry Research: Outline for Draft Strategic Plan.* ACIAR, Canberra, Australia.

Anderson, J.R. (1992) Agricultural research institutions and priorities in an era of resource scarcity: discussion. *American Journal of Agricultural Economics* 74(5), 1111–1113.

Anderson, J.R., Binkley, C.S., Fasehum, F.E., O'Hare, D. and Shepard, G. (1995) Internally Commissioned External Review. Unpublished manuscript.

Audi, P. and Mills, B. (1998) Translating farmer constraints into research themes. In: Mills, B. (ed.) *Agricultural Research Priority Setting: Information Investments for Improved Use of Research Resources.* ISNAR, The Hague, The Netherlands.

CGIAR Forestry Working Group (1991) *Research on Forestry in Developing Countries: A Proposed CGIAR Initiative.* CGIAR Secretariat, Washington, DC.

CGIAR Science Council (2008) *Impact Assessment of Policy-oriented Research in the CGIAR: Evidence and Insights from Case Studies.* A Study commissioned by the Science Council Standing Panel on Impact Assessment. CGIAR Science Council Secretariat, Rome, Italy.

CIFOR (1996) *CIFOR's Strategy for Collaborative Forestry Research.* CIFOR, Bogor, Indonesia.

CIFOR (2000) Center for International Forestry Research Strategic Priority Setting. Workshop Report. Unpublished manuscript.

Gregersen, H. and Spears, J. (1989) *Priority Forestry Research Areas with Potential for International Centralization.* CGIAR Technical Advisory Committee Secretariat, Rome, Italy.

Healy, A.T.A. (1999) Priority Setting – Building CIFOR's Capacity to Make a Difference. Unpublished manuscript.

ITFFR (1988) *A Global Research Strategy for Tropical Forestry.* International Task Force on Forestry Research, UNDP, New York.

Kelley, T. and Gregersen, H. (2003) NRM impact assessment in the CGIAR: meeting the challenge and implications for ICRISAT. In: Shiferaw, B. and Freeman, H.A. (eds) *Methods for Assessing the Impacts of Natural Resource Management Research*. A summary of the proceedings of an International Workshop, 6–7 December 2002, International Crops Research Institute for the Semi-Arid Tropics (ICRISAT), Patancheru, India.

Lele, U., Barrett, C., Eicher, C.K., Gardner, B., Gerrard, C., Kelly, L., Lesser, W., Perkins, K., Rana, S., Rukuni, M. and Spielman, D.J. (2003) *The CGIAR at 31: An Independent Meta-Evaluation of the Consultative Group on International Agricultural Research. Operations Evaluation Department*. World Bank, Washington, DC.

Maredia, M.K. and Raitzer, D.A. (2006) *CGIAR and NARS Partner Research in Sub-Saharan Africa: Evidence of Impact to date*. CGIAR Science Council, Rome, Italy.

Nilsson, G., Luckert, M.K., Armstrong, G.W., Hauer, G.K. and Messmer, M.J. (2004) Approaches to setting forestry research priorities: considering the benefits of reducing uncertainty. *The Forestry Chronicle* 80(3), 384–390.

Raitzer, D.A. (2005) Demands and Applications for Demands and Applications for Impact-Related Information in Development-Oriented International Natural Resources Management Research. Unpublished manuscript.

Raitzer, D.A. and Ryan, J.G. (2008) State of the art in impact assessment of policy-oriented international agricultural research. *Evidence and Policy* 4(1), 5–30.

SPIA (2006) *Impact Assessment of Policy Oriented Research Impact in the CGIAR: A Scoping Study Report*. CGIAR Science Council, Rome, Italy.

Spilsbury, M. (2000a) Using a 'systems' approach to research evaluation. In: *The Future of Impact Assessment in the CGIAR: Needs Constraints and Options*. Proceedings of a workshop organized by the Standing Panel on Impact Assessment of the Technical Advisory Committee. CGIAR Technical Advisory Committee Secretariat, Rome, Italy.

Spilsbury, M.J. (2000b). *DFID Support to CIFOR: Report on Activities Funded Under the DFID RLD/CIFOR Grant Agreement, April 1999 – March 2002*. CIFOR, Bogor, Indonesia.

Spilsbury, M.J. (2001) From Research to Impact – and the Tricky Part in Between. Proceedings of FAO FORSPA Workshop Communication between Forestry Researchers and Policy Makers/Stakeholders' Chiang Mai Thailand May 2001.

Spilsbury, M.J. and Kaimowitz, D. (2000) The influence of research and publications on conventional wisdom and policies affecting forests. *Unasylva* 203, 3–10.

6 Research Evaluation and Priority Assessment at the International Crops Research Institute for the Semi-arid Tropics (ICRISAT): Continuing Cycles of Learning to Improve Impacts

JUPITER NDJEUNGA AND CYNTHIA BANTILAN

Abstract

This chapter documents the research priority assessment methods used at the International Crop Research Institute for the Semi-arid Tropics (ICRISAT). Research evaluation and priority assessment have evolved to provide continuous cycles of learning to improve impacts. Prior to 1992, research priorities were established based on consultative meetings with ICRISAT and National Agricultural Research Systems (NARS) scientists to identify key productivity constraints and propose research themes and approaches to address them. For its 1994–1998 Medium-term Plan (MTP) cycle, ICRISAT undertook a quantitative priority-setting exercise using clear criteria for establishing choices among competing research activities. This drew on scientists' empirical and intuitive knowledge base. Research themes identified were impact-oriented, projecting clear milestones against which progress can be measured and evaluated *ex post*. This identification formed an integral part of the research evaluation process and facilitated revising priorities in the light of such experiences.

Following the quantitative priority assessment exercise, ICRISAT pursued extensive discussions with partners in the MTP 1998–2000 cycle and broad targets were identified to capture the areas of research and the nature of the benefits they intended to deliver. Hence, four targets were articulated by ICRISAT including prosperity, diversity, environment and inclusiveness. Due to time and cost constraints, simple scoring methods were used to rank identified constraints through a broader consultation between ICRISAT and all partners in the research and development (R&D) continuum. The target of inclusiveness included participatory methods that facilitate the participation of stakeholders and allow them to express their preferences. Scoring methods were used to rank priorities in the subsequent three-year MTP cycles.

In addition, ICRISAT enhanced priority assessment through institutionalization, building up a structured database serving as a benchmark of reference for future research evaluation, including qualitative impact indicators, using the results from numerous diffusion, adoption and impact assessment studies in setting priorities and mainstreaming poverty considerations. In recent years, CGIAR system priorities provided a framework in which to cast ICRISAT priorities. Strengths and limitations of the different methods are highlighted.

Keywords: Priority assessment, agricultural research, scoring

Introduction and Background

The International Crops Research Institute for the Semi-arid Tropics (ICRISAT) was established in 1972 with a mandate to improve sorghum, pearl millet, groundnut, chickpea and pigeon pea productivity in the semi-arid tropics. During the last 3 decades, ICRISAT expanded its research agenda along much of the research for development continuum. The goals and objectives of research have broadened from increased food production to include sustainable resource management, equity, gender, health and environmental concerns and farm- and policy-level implications. Along with the expansion of the research agenda, there is greater appreciation of the need for quantifying the economic returns to research investment and other dimensions of impact (social, environmental and institutional). In line with these changes, priority assessment in agricultural research has changed, with the principal focus shifting from yield and nutritional gains to achieving equity and environmental sustainability. This change is reflected in ICRISAT's evolving vision and strategy as well as research priorities.

The pursuit of a well-balanced and well-focused portfolio has become imperative. It has motivated stronger accountability mechanisms and systematic priority setting. Thus, the establishment of a transparent, consistent, objective and participatory priority-assessment process has become essential in institutional decision support and research planning. This process has prompted awareness among agricultural scientists and research managers about the expected benefits and pay-offs from research.

Research Priority Assessment at ICRISAT: Evolution from the 1980s to the Present

Research priority setting involves a process of explicitly or implicitly making choices over possible research activities. ICRISAT has conducted formal or informal priority assessment exercises to help set the research agenda, guide allocation of research resources and improve the quality and efficiency of research. The outcome of these exercises is a ranking of research programmes, projects or research themes within a programme or global theme. Research priorities are set across commodities, regions, disciplines, technology types and research problems. Priorities are set at different levels including the global, regional, national, research programme and project. Decisions about resource allocation also differ depending on the level at which priorities are set.

The priority assessment initiative at ICRISAT has been sustained by a determination to build an objective and transparent basis through its Medium-term Plans (MTPs). ICRISAT faced the challenge of a changing external environment where funds for research were declining, and pursuit of a focused research agenda became imperative.

Early Priority Assessment Approaches (1980s–1993)

During the late 1980s to the early 1990s, ICRISAT followed the CGIAR Technical Advisory Committee's (CGIAR/TAC) guidance, which identified four basic factors for identifying agricultural research priorities. These factors included: (i) comparative advantage (e.g. the advantage that ICRISAT has in undertaking projects where long-term, continuous effort is required); (ii) internationality (i.e. the existence of externalities and spillover effects); (iii) partnership (i.e. encouragement of inter–Center and Center–NARS activities); and (iv) efficiency and equity.

The fourth factor related to total potential benefits and high-expected payoffs, in consideration with the distributive consequences of successful research. This distributional element meant identifying the area (ecological and geographical regions) and people affected, the benefits of research in relation to costs, the feasibility of implementation and successful completion and the potential effects on the livelihoods of the poorer or marginalized sections.

Efforts to establish research priorities were based on consultative meetings with ICRISAT scientists and National Agricultural Research and Extension System (NARES) partners to identify key productivity constraints and propose research themes and approaches to address those. These exercises help to build consensus around important issues especially when ICRISAT resources were expanding. This approach did not provide information on trade-offs between various research undertakings.

Quantitative Priority Assessment for the 1994–1998 MTP

For its MTP cycle 1994–1998, ICRISAT undertook a more significant effort for research priority assessment. It involved application of a participatory approach and it set out on prioritizing among numerous competing research possibilities to make optimum use of scarce research funds against the background of a strategic plan. ICRISAT used an *ex ante* multi-objective framework, considering indicators for economic efficiency, equity, internationality and sustainability, for assessing research priorities. A supply-side methodological orientation was used to complement the TAC (1992) demand-side analysis. As illustrated later, the distinct advantage of the quantitative framework that was established is that at a time of intense competition for scarce funds, it made explicit the benefits that would flow from additional investments to an institute as well as the opportunity costs corresponding to reductions.

The priority-setting methodology used by ICRISAT (see Box 6.1) was found to provide clear criteria for establishing choices among competing research activities. It is more analytically rigorous, draws on scientists' empirical and intuitive knowledge base and is transparent and interactive. Research themes were identified along with expected impacts, projecting clear milestones against which progress can be measured and evaluated. The assumptions about prospective yield increases, research lags, probabilities of success and adoption lags and ceilings were tested against actual delivery of a new research-induced technology. This formed an integral part of the research evaluation process and facilitated revising priorities in the light of such experiences. This type of methodology was also later applied in other CGIAR Center (IRRI, 1997; ILRI, 1999).

The seminal work of Kelley *et al.* (1995) laid the groundwork for rigorous priority assessment at the institute level. The methodology was used by ICRISAT to develop its MTP 1994–1998 and is described below.

Scoring model: a weighted composite index

This methodology was developed by Kelley *et al.* (1995), and was used in setting priorities for ICRISAT's 1994–1998 MTP. It provides criteria for establishing choices among competing research activities, drawing on scientists' empirical and intuitive knowledge base. Research themes identified are impact-oriented, projecting milestones against which progress can be measured and

Box 6.1. MTP 1994–1998: Research Priority Setting Based on New 'Tandem Matrix' Management Model

The research priority-setting delivery mechanism applied a scoring method, using a weighted composite index. ICRISAT's research agenda was restructured into 22 global research projects focused on priority needs and research opportunities of the 29 production systems across the SAT. Each of the projects has a team with clearly defined objectives and milestones. The team was accountable for the development, conduct, management, resource utilization, and reporting and impact assessment of the project.

An organizational framework employing a tandem matrix was developed to facilitate the definition, development, management and conduct of projects (ICRISAT, 1992). There are two dimensions of the matrix. The original axis has four geographic regions. The geographic regions are complemented by seven disciplinary research divisions on the vertical axis that have global responsibilities. The axes of the matrix are designed to emphasize shared responsibilities, goals and outcomes through development and delivery of a relevant global research project portfolio. Research partnerships with NARS are a crucial part of the research agenda, and they have been involved in the development and conduct of these research projects.

evaluated *ex post*. The assumptions about prospective yield increases, research lags, probabilities of success and adoption lags and ceilings can be subsequently tested against actual delivery of a new research-induced technology.

A working group of scientists from across commodities, disciplines, programmes and locations was constituted to coordinate activities, exchange information and generate ideas. The group helped develop methodology and agreed upon procedures for prioritizing research themes. Four principles guided the research planning process: (i) the methodology adopted should provide clear criteria for establishing choices between competing research activities; (ii) the methodology used should be analytically rigorous and offer a consistent method for prioritization; (iii) the research plan should be based on empirical and intuitive judgements from a knowledge base within ICRISAT and NARS; and (iv) the process should be transparent and interactive with open presentations and discussions among all scientists in the organization.

These principles resulted into four major elements which characterize the process: (i) the choice of four selection criteria which reflected the mandate of the CGIAR and ICRISAT: economic efficiency, equity, internationality and sustainability; (ii) an overall score (composite index) for each potential research theme; (iii) a formalized database of primary and secondary data; and (iv) an institute-wide effort with multidisciplinary spillovers.

Relevant criteria were identified based on ICRISAT's stated mandate and objectives. A scoring method employing a weighted, additive composite index was developed to calculate an overall score for each proposed research activity or 'theme'. Subsequently, an ordinal ranking of themes and their cumulative costs emerged which defined research priorities, based on a multi-impact measure of economic efficiency, equity, internationality and sustainability. Depending on the budget available, an optimal research portfolio could be defined.

To measure economic efficiency, the expected benefit/cost ratio was estimated using a simplified version of the producer–consumer surplus approach, i.e. estimating gross welfare benefits. Equity was measured in terms of the number of poor and number of adult female illiterates in the regions where adoption is predicted. Internationality was measured using a Simpson index (Simpson, 1949) of the expected spread or spatial diversity of the problem each research theme is to address. Sustainability was measured as a subjective index from 1 to 5 in accordance with the expected contribution of the research to sustainable agriculture. These four indices were then normalized, weighted and combined into a single index (Kelley *et al*., 1995).

Key parameters used in the scoring model

Values for these measures described above were calculated using information from research protocols ('themes') and other databases. Themes were ranked from highest to lowest with respect to one specific measure. However, since themes were evaluated for impact using multi-objective criteria, some means of integrating the various measures were indispensable.

The ideal research theme has the largest expected benefits relative to the cost of investment, affects the largest number of poor, is widely pervasive and is environmentally friendly. Satisfying the two criteria for efficiency and equity means offering a high rate of return and affecting a large number of people in poverty. This preference is depicted in Fig. 6.1 as the area that should receive strong emphasis. As noted above, those with low economic returns and which affect a low number of poor and illiterate females receive limited support. In many cases, there are trade-offs, e.g. a project may affect a large number of poor, but offer only a low rate of return on investment. These types of projects may receive emphasis, depending on the social or economic objectives pursued.

The basis for assigning weights to the criteria was transparent. After considerable discussion, and given the fact that each of the four criteria was a fundamental goal of the institution, it was decided to give an equal weight to the four criteria. Therefore the model calculated the score for each theme as follows:

$$Cli = X_{li} + 0.5X_{2i} + 0.5X_{3i} + X_{4i} + X_{5i}, \quad (6.1)$$

where Cl_i is the composite score for theme i and X_{1i}, X_{2i}, X_{3i}, X_{4i} and X_{5i} are theme i normalized values for net benefit/cost ratio, number of poor, number of female illiterate, internationality and sustainability.

Although equal weights are applied in the scoring model, the trade-offs between the different criteria are not equal (Appendix 6.1). In order to achieve a 0.25 increase in the composite index score, a research theme would have to either: (i) increase its net benefit/cost ratio by 12.5; (ii) increase the number of poor affected by 62.5 million and the number of female illiterates affected by 75 million; (iii) increase the Simpson index of diversity by 0.25; (iv) increase the sustainability score by 1; or (v) increase any fractional combination thereof. The composite index is indifferent between those changes.

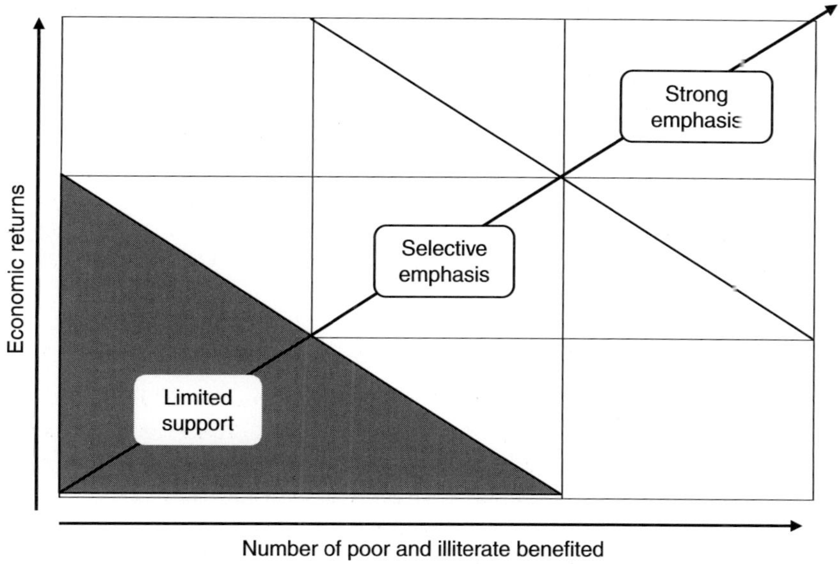

Fig. 6.1. Research portfolio selection: potential trade-offs between returns and equity.

Developing the plan and data requirements

In developing the research plan, the steps in conceptualizing, operationalizing and quantifying proceeded in an iterative fashion. These steps included everything from adoption domain definition to research theme development, and from the specification of scientific outputs to expected economic and social impact. The first task in developing the plan was to define the research domain, which was essential in assessing potential regional research impacts and spillovers. Production systems and yield constraints were the primary criteria for defining adoption domains.

Data requirements

Once the adoption domains were identified for each constraint, data were gathered on population, absolute numbers of poor, female literacy rate, crop production, crop production value, crop losses per constraint, economic losses, and yield and potential economic recovery through research.

Socio-economic data were derived from various sources including the World Bank (1992), UNDP (1992) and Broca and Oram (1991). World prices are used to assess the value of crop production gains and losses. Production data were obtained from the FAO (1992) and crop reports from state and regional levels of India. Crop yield losses resulting from biotic and abiotic constraints were elicited from scientists or estimated from survey results. Estimates were expressed as a percentage of average yields currently achieved in the respective country. Yield recovery or yield gain was estimated as a portion of the respective contributions from resolving each single constraint.

Research protocols

After identifying the adoption domains, constraints and associated yield losses were translated into research themes and expected outputs. From the many constraints identified, a limited number of research themes were identified. For each research theme, a research protocol was identified and generated by the respective scientists. The protocols covered 14 items which provide qualitative and quantitative information: constraint/problem, crop(s) involved, research domain, type of research, research output, research and extension lags, adoption ceilings, probability of success, senior scientist years, yield improvement expected, production costs, stability component, environmental/sustainability component and extra-capital requirement.

Measuring impact

Economic value of research

The unadjusted and undiscounted economic benefits are estimated as:

$$EB_{ikt} = [(y_{ikg}{}^{*}Pkg) + (y_{ikf}{}^{*}P_{kf}) - c_{ikt}]A_{kt}, \tag{6.2}$$

where EB_{ikt} is the undiscounted economic benefits derived from new technology i on crop k in year t, y_{ikg} and y_{ikf} are the per hectare increment (average) in grain and fodder yield improvement expected on farmers' fields due to adoption of technology i, P_{kg} and P_{kf} their respective prices, c_{ikt} is the per hectare incremental costs associated with adoption of technology i and A_{kt} is the total area of crop k in the relevant research domain.

For each of the research theme, the economic benefits are described as the net present value (NPV) of the gross economic benefits derived from technology i on crop k in year t (NPV_{ikt}) where Pr_i is the probability of success in research leading to the development of technology i, Ad_{it} is the percentage of adoption of technology i by farmers in year t, and r is the social discount rate.

$$NPV_{ikt} = \sum_{t=1}^{n} \frac{EB_{ikt} \times PR_i \times Ad_{it}}{(1+r)^t}. \qquad (6.3)$$

This value could not be attributed solely to the research done by ICRISAT because of the contribution of other partners in the R&D continuum, such as NARS, public and private seed companies, agricultural extension and infrastructure development. Benefits were estimated as gross benefits, and investment costs were not subtracted. The single largest cost is human capital (scientists, support staff and operational costs). The discounted stream of cost flow for technology i and crop k are calculated as:

$$C_{ik} = \sum_{t=1}^{n} \frac{(SY_t \times SC) + K_t}{(1+r)^t}, \qquad (6.4)$$

where C_{ik} is the discounted costs associated with technology i on crop k, SY_t is the total number of scientist years in year t, SC standard scientist costs per year, K_t is the special capital costs in year t, $t = 1, 2, \ldots, n$; r represents the discount rate.

The net benefit/cost is the ratio between the gross benefit (Eq. 6.4) and the costs (Eq. 6.5) is

$$\text{Net } B/C \text{ ratio} = \frac{NPV_{ik}}{C_{ik}}. \qquad (6.5)$$

Equity

A combination of two poverty measures: (i) head count of poor, defined as the number of people with income below the minimum level required to maintain an acceptable calorie consumption level; and (ii) the extent of female illiteracy were used as proxies for general welfare. Data on the absolute number of poor were taken from an International Food Policy Research Institute (IFPRI) study by Broca and Oram (1991); the number of adult female illiterate from UNDP (1992), Vu (1984), UNESCO (1985) and Indian census statistics (Govt of India, 1985).

Internationality

The cross national character of a research theme was considered as a prominent feature in determining an international research institute's priorities. As a measure of internationality, the Simpson index of diversity was used:

$$D_i = 1 - \sum_{j=1}^{J} (S_{ij}/100)^2, \tag{6.6}$$

where D_i represents the diversity value ranging from 0 to 1, and S_{ij} is the share of the total production gain resulting from research theme i in country j. Higher values correspond to research themes which are more international in scope.

Sustainability

The contribution of the research theme to sustainability is difficult to measure. As objective valuation is challenging with present methods, we used a subjective scale. Themes are ranked from 1 to 5 according to the likelihood impact in maintaining or upgrading the resource base to ensure long-term productivity. A 1 indicates no, or only negligible, contribution to sustainable agriculture; a 5 indicates upgrading the resource base is the primary focus of the research. No rigorous formula is used but arguments are put forward to explain how a particular piece of research is expected to change the resource base that in turn would impact long-term productivity growth.

Consolidation, deletion and ranking

For each of the research themes, an overall score was computed using the weighted, additive composite index scoring model described earlier. Subsequently, an ordinal ranking of themes emerged.

Prior to final ranking and before the scoring exercise, senior research management scrutinized all themes for consolidation and elimination, as there were research themes for which the institute does not have a comparative advantage. After this early screening, the final consolidated research themes were developed to establish priorities for the Institute.

The research portfolio plan

The ranked research themes and their estimated costs for ICRISAT's MTP exercise are given in Appendix 6.1. The final research portfolio largely depends on final budget allocation. Totally 110 research projects were selected, several of which could not have been included if there had not been special funding. The results also illustrated the opportunity costs of budget cuts at the margin. For example, if the budget decreases from a US$19.7 million to US$19.0 million, the foregone benefits would be those for research themes 55–57. The NPV from these three themes was US$35.7 million which represented the potential benefit stream lost if an additional investment of US$0.70 million for core budget was not made available to ICRISAT.

Scoring Approaches from 1999–2007

In the follow-up MTP cycle for 1998–2000, ICRISAT pursued extensive discussions with partners in which broad targets were identified that captured the areas of research and the nature of the benefits they intended to deliver through these partnerships during the MTP period. Four scoring criteria used by ICRISAT include prosperity, diversity, environment and inclusiveness.

1. *Prosperity.* Poverty is a fundamental cause of hunger, disease, environmental degradation and a host of other afflictions. Since the majority of the poor in the Semi-arid Tropics (SAT) are engaged in farming or other agriculturally related enterprises, the road to prosperity lies in the development of more productive and efficient agricultural systems.
2. *Diversity.* Poor farmers with small landholdings cannot afford the risk of being overly dependent on just a few crops or cropping systems. Diversity creates options; it spreads risk; it evens out peaks and valleys in labour use and income; and it enables the creation of added value by expanding the application of farmers' management skills to new enterprises. More diverse, complex cropping systems are usually more robust and stable, and sustainable over time.
3. *Environment.* Environmental resources are the fundamental inputs of agriculture. The conscious or unconscious abuse of these resources can throw entire societies into poverty. This target has particular relevance to the SAT where poverty is a driving force behind short-term exploitation of the environment to satisfy pressing food needs.
4. *Inclusiveness.* Research products must be understood and valued by those who use them if they are to have impact. It is difficult to achieve this unless stakeholders are involved in the identification of relevant research priorities, and in the research process itself.

The target of inclusiveness appealed to participatory methods to support the priority-setting process and decision-support tools that facilitate the participation of stakeholders, allowing them to express their preferences.

Subsequent three-year MTP cycles followed, and the criteria used to rank priorities were more or less maintained. The criteria were broadened to include equity, efficiency, internationality sustainability, new science opportunity, relevance to NARS priorities and future trends (Deb and Bantilan, 2001). The strategies and priority guidelines offered by the CGIAR TAC (later called Science Council) were influential in this evolution. Notably, major efforts continue to be launched to consult NARS partners and other stakeholders in the setting of priorities. The approaches to strategic planning and priority assessment in the CGIAR continued to advance in the last few years, where the basis of priority assessment not only became more inclusive and participatory, but also increasingly appealed to process plans for strategic planning, impact pathways, situation and outlook analysis, periodic commodity and sector reviews, and more systematic understanding and foresight of the external environment and mega trends (see Box 6.2). These

Box 6.2. ICRISAT: Understanding the External Environment – SAT Futures Approach

The agricultural environment in the SAT is constantly changing, in terms of cropping patterns, income opportunities, trade regulations and other factors. In order to remain relevant, ICRISAT monitors these changes and their implications for priorities of its research agenda. This monitoring process was formalized as a global research theme in the early 2000s (one of six themes at ICRISAT at that time) titled *SAT Futures and Development Pathways*. This global theme has three broad objectives: (i) to track changes in the external environment and better understand the factors driving these changes; (ii) correspondingly, to review (and adjust where needed) ICRISAT's research agenda, priorities and funding allocations among alternative research areas; and (iii) to provide an analytic, objective basis for research management decisions, i.e. a decision-support system for senior management.

The SAT Futures project includes strategic socio-economic research in specific areas: commodity trends and market outlooks; input supply and access constraints; patterns and determinants of technology adoption; institutional innovations; and dynamics and determinants of poverty. These studies identify technological, policy and institutional alternatives and development pathways to enhance the livelihoods of smallholder farmers in the SAT. They also inform and direct ICRISAT's research investment towards the most crucial areas.

The project uses a participatory approach. ICRISAT organized a series of brainstorming meetings to discuss poverty-related problems and their implications for research priorities. Many key stakeholders are involved to ensure that the final outputs reflect a diversity of views and experiences. These include national and international institutes, development investors, universities, the private sector, extension, non-governmental organizations (NGOs) and farmer organizations. The broad involvement also enables tapping of a large, multidisciplinary pool of expertise in policy and planning, sustainable development, rain-fed agriculture, agricultural economics, farming systems research, germplasm enhancement and environmental conservation. Simultaneously, focus group meetings were also conducted in each region (East Africa, West and Central Africa, Southern Africa, South Asia, South-east Asia), involving scientists from ICRISAT and partner institutions.

The SAT Futures approach in strategic planning and priority setting follows a systematic procedure: literature survey, data analysis, stakeholder consultations and synthesis of the major issues. It seeks to identify the unique features of the SAT, and understand the differences in agricultural trends between the SAT and other regions of the developing world. During ICRISAT's 2006 research priority setting and visioning exercise, the process was supported by a review of major trends in SAT agriculture using available long-term time series data from the 1960s. The review summarized the major constraints limiting income growth, poverty alleviation, food security and environmental sustainability now and towards 2020, the implications for future R&D strategies and priorities for the SAT and the roles for ICRISAT, NARS, NGOs and the private sector in implementing these R&D strategies.

In sum, these initiatives led to: (i) development of guidelines to enhance participation in research; (ii) clear identification of key issues and external factors affecting SAT agriculture, emerging challenges and opportunities and strengths as well as gaps in existing research systems; (iii) documentation: synthesis report summarizing responses from the baseline survey, as well as collation of relevant literature from other sources (e.g. World Bank and FAO); (iv) development of a framework that underpins the critical issues in SAT agriculture, linking productivity, food security and poverty reduction; (v) update and analysis of micro-level data and macro-level statistics (both demographic and agricultural) to support research decision making; (vi) design of research for development strategies for the SAT; and (vii) publication of several important strategy documents: (a) Future challenges and opportunities for agricultural R&D in the SAT; (b) Future of agriculture in African SAT; and (c) Vision on SAT agriculture for Asia.

Continued

Box 6.2. Continued

The SAT Futures project, too, has evolved, in response to this consultative exercise. Research has been refocused on three areas: (i) strategic assessments for agriculture and economic growth in the SAT of Asia and Africa and implications for agricultural research priorities; (ii) development pathways and policies for rural livelihoods; and (iii) synthesis studies: lessons learnt from impact studies, institutional arrangements and implications for research spillovers across regions.

The key question is: 'How can agricultural research improve the pay-offs to diverse and changing investment opportunities?' The ultimate objective is to steer research direction and development towards a pathway that better addresses poverty and environmental degradation.

approaches shaped the current ICRISAT research priority assessment effort, and were found to be consistent with the current CGIAR research priorities and regional research priority assessment by regional bodies such as West and Central African Council for Agricultural Research and Development (CORAF/WECARD), Association for Strengthening Agricultural Research in Eastern and Central Africa (ASARECA) and Southern African Development Community (SADC).

In particular, with increasing pressure resulting from dwindling core funding in the period from 1999 to 2002, time and funds involved in implementing the rigorous priority assessment approach used in the 1994–1998 cycle were not available and hence a simpler scoring method was used. In this case, a range of constraints was first identified through a large consultation process with partners in the R&D continuum through a survey of significant constraints and opportunities in the semi-arid tropics. The results were shared in a series of consultative meetings at the regional level in West and Central Africa, Southern and Eastern Africa and Asia. During these meetings, constraints and opportunities were translated into research themes. Using a matrix format, a simple mean-scoring method was used to rank the relative importance of these research themes.

Continuing Enhancements to Priority Assessment at ICRISAT

Structured database

Systematic calculation of the measures of the various priority-setting criteria requires a structured database. The database developed from the research evaluation and impact assessment (REIA) project of ICRISAT contains comprehensive information on variables including research objectives, target research domains, estimated yield losses, expected yield gains, probability of success, adoption rates and ceiling levels, research and adoption lags, expected output and manpower and capital requirements. This database serves as a benchmark or reference for research evaluation of future projects. This database is regularly updated through impact monitoring.

Institutionalization

ICRISAT research management instituted a continuous cycle of priority assessment with a defined and regular interval to provide an avenue of feedback and timely redirection of research. Establishing such a mechanism (Joshi and Bantilan, 2001) required the following essential steps: (i) adaptation of a uniform methodological framework to assure comparability and consistency of identified priorities; (ii) regular annual database updates; (iii) establishment of a monitoring process for performance, adoption and impact; and (iv) training to develop the capacity of scientists associated with priority assessment. Training is essential not only to undertake priority assessment consistently and objectively, but also to achieve transparency and active participation within the organization. Finally, in order to facilitate organizational priority-setting processes, a monitoring and evaluation (M&E) scheme was institutionalized by ICRISAT Governing Board in 2004, whereby M&E should be written into research proposals such that movement along the research evaluation and impact pathway continuum can be monitored, any necessary mid-course adjustments could be made and *ex post* impact assessments properly done.

Inclusion of qualitative impacts in priority assessment

Since research evaluation and priority assessment involve the process of making choices in the context of scarcity, most of the earlier efforts have placed emphasis on the economic principles of efficiency and on the costs and benefits that can be expressed in monetary values. The latter raised concerns about qualitative aspects associated with externalities (Bantilan and Davis, 1991; Brennan and Bantilan, 2003), gender and distributional effects and longer-term impacts which tend to be neglected with such an emphasis.

For inclusion of qualitative impacts in priority assessment, a systematic documentation of the impact pathways has been useful in identifying the sources of the qualitative effects of technology adoption (Bantilan *et al.*, 2005). The pathway helps in clarifying the nature of impacts by considering whether or not the expected changes due to technology adoption can be valued using conventional markets, and therefore identifying variables that have market impacts and those that relate to non-market effects. A listing of the potential positive and negative effects aids in the analysis of the market and non-market impacts of alternative technology options. This analysis is particularly useful for assessing qualitative effects and relative preferences among alternatives. It records the market impacts reflecting yield gains or reduced yield losses and changes in unit cost. The measurement of environmental effects in monetary terms within the context of economic surplus draws from changes in the social marginal cost of production (i.e. product supply) and the demand for the marketed product. The inventory of non-market effects may be substantial, e.g. significant positive effects may result in longer-term yield stability, or increased resource availability in the future. A detailed account of the analysis of possible market and non-market impacts is presented in Bantilan *et al.* (2005). This study explains how conventional calculations that exclude environmental effects can skew measures of the full potential benefits from an improved technology.

Using the Results of *Ex post* Impact Assessment in Priority Assessment: Learning Cycles and Feedback Process

The process of assessing impact *ex post* can generate insights that can help to better inform *ex ante* priority assessment and provide grounds for additional investment in the resultant research portfolio (Bantilan and Ryan, 1996). For example, data from primary field studies provide a good basis for reasonable estimates of parameters, which are used in the priority-setting exercise, such as: (i) levels and speed of adoption and reasons for non-adoption of technology; (ii) farmers' perceptions of desirable traits or features of technology options; (iii) on-farm gains due to alleviation of biotic and abiotic constraints; and (iv) infrastructural, institutional and policy constraints in facilitating technology exchange.

Two categories of impact data may be developed. The first is a set of primary data on adoption and related variables generated from formal and informal on-farm surveys. The second is a set of secondary data based on input from partner agencies. On-farm reconnaissance and formal surveys may be primarily aimed at continuously assessing the extent of adoption of improved technology from the secondary database. This confirms the extent of utilization of improved technologies by farmers in the target regions. Research lag is a major parameter determining the present value of research, and the cost of miscalculating it in terms of erroneous priority ranking can be substantial. Verification of research and adoption lags used can be accomplished by cross-checking *ex post* data from various sources.

Farmers' opinions on important constraints as well as their perceptions of desirable cultivation and management technology options may also be generated from primary surveys. These farmers' perspectives provide the following information: (i) they identify the constraints and research opportunities; (ii) they provide an empirical basis for the expected ceiling levels of adoption, i.e. technologies introduced in an environment characterized by significant bottlenecks to adoption cannot be expected to have high adoption ceilings unless these constraints are addressed; and (iii) they identify the research options that directly address the users' needs and are most likely to be adopted.

Estimates of yield losses due to important constraints and on-farm gains due to improved technology are also vital pieces of information for deciding research priorities. Impact studies can be used to validate estimates of expected yields. Furthermore, the estimates generated from these surveys (i.e. yield gains or unit cost reductions) also provide a way of predicting the potential supply shift, a necessary parameter for estimating potential impacts in cost/benefit analyses.

Another important outcome from impact studies is the assessment of researchers' perceptions or constraints, which can be technological, institutional, infrastructural and policy-based. Two aspects are relevant for seed policy and priority assessment: (i) standard variety release procedures of breeders' selection of materials that can make it through the formal release system; and (ii) criteria for varietal release do not necessarily match farmers' needs and preferences.

Mainstreaming Poverty Considerations in Priority Assessment

Mainstreaming poverty considerations is an important issue in ICRISAT priority assessment (Bantilan and Keatinge, 2007) in the light of recent developments in the global research agendas of international organizations, which have identified poverty eradication as a common goal (UN, 2002; CGIAR, 2005). Mainstreaming poverty recognizes that there are at least five ways by which agricultural research can benefit the poor: (i) increasing poor farmers' productivity; (ii) greater agricultural employment opportunities for small farmers and landless workers; (iii) higher wages and growth in adopting regions; (iv) lowering food prices; and (v) greater access to nutritive crops.

Ryan (2004) identified the following considerations in relation to poverty-targeted agricultural research priority assessment: (i) it is not necessarily given that research investments targeted at the locations of the poor will achieve maximum impact on the resident poor. Many factors mediate this relationship and make it difficult to argue that priorities at the macro level should be primarily based upon the location of the poor. These factors include price effects, migration and research spillovers in other regions. However, as Fan and Hazell (2000) have shown, the marginal returns to research are higher in less-favoured environments and also the effect of this on poverty alleviation is greater. Therefore it is not clear that it is appropriate to neglect the less-favoured areas and allow 'trickle down' forces from more favoured areas to equilibrate the benefits; (ii) wage and employment effects of targeted research can be counter-intuitive. In particular, if labour-intensive commodities have non-responsive demands, then research on those commodities could lead to mechanization or to their substitution in production of less labour-intensive commodities; and (iii) growth linkages between agricultural and nearby rural industry can generate significant multiplier effects, benefiting the poor most when agricultural income is a high proportion of total income.

By analysing a typology of agricultural regions based upon agroecological zones and socio-economic factors that condition the size and distribution of benefits from technological change, five broad areas of focus for a pro-poor research agenda have been identified (Haddad and Hazell, 2001): (i) increasing productivity in less-favoured lands, especially in heavily populated areas but also in high-potential lands constrained by poor infrastructure and market access; (ii) increasing production of staple food in areas where food price effects are still important and/or in areas that have a comparative advantage in growing these crops; (iii) helping smallholder farms to diversify into higher-value products, especially in areas where market prospects are good; (iv) increasing employment and income-earning opportunities for landless and near-landless workers in labour-surplus regions; and (v) nutritional enhancement of diets by investing in agricultural technology that reduces the price of micronutrient-rich foods; increases physical access in remote rural areas, or increases the nutrient content of food staple crops via traditional or transgenic technologies.

Reflections and Ways Forward

The environment facing publicly funded international agricultural research centers such as ICRISAT has changed significantly over the past two decades (Byerlee, 2000; ICRISAT, 2002). One important dimension of the changing agricultural research environment is the increased emphasis that is now given to food security and poverty alleviation. The CGIAR explicitly recognizes that investments in international public goods-oriented research must have poverty and impact foci (GFAR & iSC, 2002). ICRISAT has embraced this perspective and is renewing its research efforts to give greater priority to problem-based, impact-driven science and output delivery (ICRISAT, 2002).

This changing context of agricultural research also implies that approaches to setting research priorities at ICRISAT must adapt to these changes. The 2002 ICRISAT vision and strategy internalizes these changes. Recent CGIAR emphasis on a regional approach to agricultural research in order to better address poverty, food security and the environment in developing countries is being assumed by ICRISAT through its regionalized research and its administrative empowerment of the regions.

The ICRISAT Vision and Strategy (ICRISAT, 2002) as well as the 2003–2005 MTP have adapted to the CGIAR's new vision, and ICRISAT has initiated steps to institutionalize regional research planning and priority assessment. ICRISAT's current research strategy is addressed in four global themes and implemented through regional projects that are based on strategic regional priorities. However, to implement regional consultative priority assessment, exercises must be done systematically at a regional level to support planning and resource allocation decisions.

Currently, the CGIAR system priorities identified by the CGIAR Science Council (described in Chapter 12, this volume) provide a framework for 80% of ICRISAT's priorities. Precedence models are used to allocate resources. In effect, the level of funding in the previous year is the basis for the following year's allocation of resources to project themes and projects. Research resources are increased or decreased marginally depending on the overall funding situation. Changes in total resources available are usually allocated in equal proportion across research themes. This approach is simple and quick and has minimal data requirements. It can also provide long-term continuity in funding of research themes and projects. One disadvantage of this model is that it can continue allocating resources to research areas that have reached the limits of their productivity and for which changing research environment mean that they are not even high-priority activities anymore. Precedence models are also not forward looking since funding decisions are based on past levels of resource allocation rather than on research investments that are likely to give the greatest impact. It is therefore difficult to use this model for introduction of new research areas.

Conclusions

ICRISAT has pursued a range of priority assessment exercises over the last three decades. The 'scoring model approach: weighted composite index' to setting priorities has its advantages. It made explicit the benefits that would flow from additional investment to the institute as well as the opportunity costs corresponding to specific reductions in research funds. It generated milestones by which research outputs can be evaluated *ex post*. However, it involved significant time and financial costs. The degree of scientific subjectivity is also significant with this approach. In addition, scientists and managers expressed concerns about the lack of simple and transparent procedures for priority assessment and resource allocation. However, there is a need for a certain level of scientific rigour to priority assessment, so as to incorporate basic economic principles. Thus, efforts to search for simple ways to prioritize research themes and resource allocation at ICRISAT continue.

Regional priority assessment is currently pursued and appears to be more appealing to partners, given the participatory nature of priority assessment across regions. However, the need for clarity over the linkages between priority assessment, research planning (including M&E), and budgeting and resource allocation continues to be a big challenge.

References

Bantilan, M.C.S. and Davis, J.S. (1991) Across-commodity Spillover Effects of Research and Opportunity Costs in a Multi-product Production Environment. ACIAR/ISNAR Project Paper No. 30. Australian Center for International Agricultural Research, Canberra, Australia/ International Service for National Agricultural Research, The Hague, The Netherlands, 24.

Bantilan, M.C.S. and Keatinge, J.D.H. (2007) Considerations for determining research priorities: learning cycles and impact pathways. Chapter 3. In: *Agricultural Research Management*. Springer, 32.

Bantilan, M.C.S. and Ryan, J.G. (1996) Using Impact Assessment in Research Priority Setting: Developments at ICRISAT. Paper presented at the International Conference on Global assessment for science policy, 26–28 August 1996, Melbourne, Australia.

Bantilan, M.C.S., Anupama, K.V. and Joshi, P.K. (2005) Assessing economic and environmental impacts of NRM technologies: an empirical application using economic surplus approach. Chapter 11. In: Shiferaw, B., Freeman, H.A. and Swinton, S.M. (eds) *Natural Resource Management in Agriculture: Methods for Assessing Economic and Environmental Impacts*. CAB International, Wallingford, UK.

Brennan, J.P. and Bantilan, M.C.S. (2003) Price and yield effects of spillovers in international agricultural research: evidence from ICRISAT and Australia. *Agricultural Economics* 28(2), 87–97.

Broca, S.A. and Oram, P. (1991) Study on the Location of the Poor. Paper Presented for the Standing Committee for Priorities and Strategies of TAC/CGIAR. IFPRI, Washington, DC.

Byerlee, D. (2000) Targeting poverty alleviation in priority setting for agricultural research. *Food Policy* 25, 442.

CGIAR (Consultative Group on International Agricultural Research) (2005) *Agricultural Research Matters – Achieving the Millennium Development Goals*. CGIAR, Washington, DC. Available at: http://www.cgiar.org/pdf/cgiar_mdgs_brochure_2005.pdf

Deb, U.K. and Bantilan, M.C.S. (2001) Spillover Impacts of Agricultural Research: A Review of Studies. Socioeconomics and Working Paper Series No. 8. International Crops Research Institute for the Semi-arid Tropics Patancheru 502 324, Andhra Pradesh, India. 32.

Fan, S. and Hazell, P. (2000) Should developing countries invest more in less-favoured areas? an empirical analysis of rural India. *Economic and Political Weekly* 35(17), 1455–1464.

FAO (1992) *Production Yearbook 1991*. FAO, Rome, Italy.

GFAR and iSC (2002) Regional Priority Setting. Agenda Item During the Stakeholder Meeting, 30–31 October 2002. Annual General Meeting, 2002. Manilla, The Phillippines. 180.

Govt of India (1985) *General Population Tables. Census of India 1981*. Registar General and Census Commissioner, New Delhi, India.

Haddad, L. and Hazell, P. (2001) CGIAR Research and Poverty Reduction. Paper prepared for the Technical Advisory Committee of the CGIAR. International Food Policy Research Institute, Washington, DC.

ICRISAT (International Crops Research Institute for the Semi-arid Tropics) (2002) *ICRISAT Vision and Strategy to 2010*. ICRISAT, Patancheru 502 324, Andhra Pradesh, India.

ILRI (International Live Stock Research Institute) (1999) *Making the Livestock Revolution Work for the Poor*. ILRI, Nairobi, Kenya.

IRRI (International Rice Research Institute) (1997) *Annual Report 1995–1996: Listening to the Farmers*. IRRI, Metro Manila, The Philippines.

Joshi, P.K. and Bantilan, M.C.S. (2001) Institutionalization of research prioritization, monitoring and evaluation. In: Suresh, P. and Joshi, P.K. (eds) *Priority Setting, Monitoring and Evaluation in Agricultural Research*. National Center for Agricultural Economics and Policy Research, New Delhi, India, pp. 65–81.

Kelley, T.G., Ryan, J.G. and Patel, B.K. (1995) Applied participatory priority setting in international agricultural research: making trade-offs transparent and explicit. *Agricultural Systems* 49(2), 177–216.

Ryan, J. (2004) Agricultural Research and Poverty Alleviation: Some International Perspectives. ACIAR Working Paper no. 56, Australian Center for International Agricultural Research, Canberra, Australia.

Simpson, E.H. (1949) Measurement of diversity. *Nature* 163, 688.

TAC (Technical Advisory Committee) (1992) *A Review of CGIAR Priorities and Strategies* (Advanced Draft – Rev. I). CGIAR Technical Advisory Committee Secretariat, FAO United Nations, Rome, Italy.

UN (United Nations) (2002) *The Millennium Development Goals and the United Nations Role*. UN, New York. Available at: http://www.un.org/millenniumgoals/MDGs-FACTSHEET1.pdf

UNESCO (United Nations Educational Scientific and Cultural Organization) (1985) *Statistical Yearbook 1985*. United Nations Educational Scientific and Cultural Organization, Paris.

UNDP (United Nations Development Programs) (1992) *Human Development Report 1992*.

Vu, M.T. (1984) *World Population Projections*. The World Bank, Washington, DC.

World Bank (1992) *World Bank Development Report 1992. Development and Environment*. Oxford University Press, Oxford.

Appendix 6.1

Example of research themes ranked by composite index. (From Kelley *et al.,* 1995.)

			Efficiency			Research cost		Equity					
Rank	Program	Constraint/theme	NPY ($mil)	Net B/C (ratio)	IRR (%)	First year ($mill)	Average 94–98 ($mill)	Poverty (million poor)	Gender (million fem. ill.)	Inter-nationality (index)	Sustain-ability (index)	Composite index	Cumulative cost ($mill)
Core	funding												
1	GRU	Germplasm evaluation	79.1	101.9	–	0.19	0.13	397.0	378.0	1.00	4	5.21	0.19
2	GRU	Germplasm collection	24.5	40.7	–	0.14	0.12	397.0	378.0	1.00	5	4.24	0.33
3	LGM	Drought-CP	265.2	113.7	47.3	0.48	0.42	60.6	119.5	0.55	3	3.64	0.81
4	GRU	Germplasm maintenance	15.3	35.5	–	0.10	0.09	397.0	378.0	1.00	3	3.63	0.91
5	LGM	Ascochyta blight-CP	73.2	134.7	64.0	0.14	0.09	9.9	74.4	0.57	1	3.48	1.05
6	LGM	Rust-GN	80.9	47.9	33.0	0.33	0.28	337.0	310.0	0.70	3	3.35	1.38
7	LGM	Aflatoxin-GN	7.6	23.1	29.3	0.05	0.04	248.2	298.6	0.82	5	3.28	1.43
8	LGM	Late leaf spot-GN	32.7	12.4	24.9	0.43	0.36	329.0	302.0	0.84	4	3.00	1.86
9	LGM	Aflatoxin (MGT)-GN	19.7	6.4	23	0.56	0.47	360.0	308.0	0.60	5	2.96	2.42
10	LGM	Insect damage-CP	76.1	78.5	47.2	0.25	0.16	88.2	107.9	0.26	4	2.94	2.67
11	LGM	Wilt-CP	63.9	114.2	57.0	0.14	0.09	88.2	107.9	0.26	1	2.90	2.81
12	RMP	Adopt. asses./imp. evl.	–	–	–	0.62	0.52	75.9	114.1	0.00	3	–	3.43
13	RMP	Res. resource. alloc'n	–	–	–	0.21	0.11	397.0	378.0	1.00	4	–	3.64
14	RMP	Soil nutrients	130.3	35.9	43.4	0.54	0.45	167.9	162.2	0.49	5	2.81	4.18
15	LGM	Early leaf spot-GN	9.1	4.4	21.3	0.45	0.37	345.0	313.0	0.70	4	2.75	4.63
16	LGM	Genetic poten'l yld-PP	64.0	63.5	41.5	0.13	0.11	125.2	168.2	0.23	3	2.53	4.76
17	LGM	Yield potential-GN	29.6	12.3	23.4	0.44	0.37	234.2	363.4	0.71	3	2.53	5.20
18	CRL	Striga-SG	78.7	41.4	46.2	0.28	0.23	31.5	43.8	0.80	4	2.51	5.48
19	LGM	Drought-GN	14.5	5.2	20.3	0.50	0.42	331.8	326.0	0.62	3	2.43	5.98
20	LGM	Root rots-CP	41.3	70.3	49.1	0.14	0.09	88.2	107.9	0.33	2	2.34	6.12
21	LGM	Bud necrosis virus-GN	1.0	1.2	13.9	0.13	0.11	298.9	328.1	0.66	3	2.33	6.25
22	CRL	Grain & stover yld.-SG	47.1	16.6	31.7	0.68	0.57	180.8	169.2	0.85	3	2.33	6.93
23	RMP	Soil fertility	86.4	21.1	29.1	0.58	0.48	16.8	37.9	0.76	5	2.28	7.51
24	LGM	St. mosaic/Fu. wilt-PP	58.5	40.4	33.5	0.21	0.17	125.2	168.2	0.12	4	2.21	7.72

25	RMP	Soil structure	29.4	5.9	22.8	0.74	0.62	167.9	162.2	0.46	5	2.18	8.46
26	LGM	Leaf miner-GN	5.7	6.0	20.8	0.19	0.16	195.7	268.6	0.46	4	2.17	8.65
27	LGM	Biolog. N fixation-CP	9.6	16.6	30.4	0.10	0.09	88.2	133.7	0.43	5	2.16	8.75
28	LGM	Leaf miner (MGT)-GN	4.8	4.5	21.3	0.23	0.19	195.7	268.6	0.46	4	2.14	8.98
29	RMP	Water deficit	122.8	19.1	32.9	0.95	0.79	154.4	151.4	0.34	4	2.03	9.93
30	LGM	Spodoptera-GN	0.7	0.9	13.3	0.14	0.12	174.7	247.6	0.40	4	1.93	10.07
31	LGM	Peanut clump virus-GN	5.7	4.9	21.0	0.23	0.19	114.3	124.0	0.84	3	1.87	10.30
32	LGM	Posette virus-GN	20.8	8.6	23.1	0.53	0.39	71.9	71.4	0.89	3	1.82	10.83
33	LGM	Helicoverpa (MGT)-PP	26.0	23.8	29.7	0.17	0.14	98.2	136.4	0.17	4	1.82	11.00
34	CRL	Stem borer-SG	8.4	1.6	16.1	0.76	0.63	232.7	191.2	0.75	2	1.82	11.76
35	CRL	Grain mold-SG	66.0	21.5	32.2	0.45	0.38	51.2	57.2	0.68	3	1.81	12.21
36	LGM	Millipedes-GN	3.0	8.0	23.8	0.04	0.03	27.3	37.2	0.77	4	1.80	12.25
37	RMP	Water deficit-PM,SG,GN	22.0	3.9	19.4	0.83	0.69	24.1	42.6	0.76	4	1.71	13.08
38	RMP	Tech. adopt./imp/eval.	–	–	–	0.29	0.24	24.1	42.6	0.83	2	–	13.37
39	RMP	Agroforestry	16.7	3.5	17.7	0.60	0.55	24.1	42.6	0.76	4	1.70	13.97
40	RMP	Char'n of prod'n envi't	–	–	–	0.72	0.60	24.1	42.6	0.76	3	–	14.69
41	LGM	Nematodes-GN,PP,CP	15.1	5.9	21.3	0.41	0.34	179.7	263.9	0.27	3	1.69	15.10
42	LGM	Termites-GN	2.3	2.4	16.7	0.11	0.09	27.3	37.2	0.77	4	1.68	15.21
43	LGM	Suboptimal yield-CP	0.9	0.5	12.6	0.25	0.21	88.2	133.7	0.52	4	1.68	15.46
44	CRL	Low temperature-SG	1.6	9.6	13.4	0.19	0.17	32.7	11.8	0.60	4	1.63	15.65
45	LGM	White grubs-GN	1.3	1.6	15.8	0.11	0.09	27.3	37.2	0.72	4	1.62	15.76
46	CRL	Head bug-SG	12.8	7.1	24.7	0.27	0.22	43.3	74.8	0.76	3	1.61	16.03
47	LGM	Drought-PP	19.7	7.7	24.0	0.41	0.35	98.2	136.4	0.28	4	1.61	16.44
48	CRL	Anthracnose-SG	13.5	4.6	25.5	0.43	0.36	126.7	110.8	0.82	2	1.60	16.87
49	CRL	Midge-SG	14.4	4.1	19.4	0.52	0.43	56.6	47.1	0.82	3	1.59	17.39
50	RMP	Char'zation of environ.	–	–	–	0.25	0.21	75.9	114.1	0.00	3	–	17.64
51	RMP	Microecon studies	–	–	–	0.41	0.41	–	–	–	–	–	18.05
52	RMP	Natural resources	–	–	–	0.60	0.50	75.9	114.1	0.00	5	–	18.65
53	RMP	Supply & demand	–	–	–	0.21	0.17	75.9	114.1	0.00	4	–	18.86
54	RMP	Farmers' preferences	–	–	–	0.14	0.12	75.9	114.1	0.00	3	–	19.00
55	RMP	Beneficial organisms	27.3	11.3	27.4	0.41	0.34	62.4	104.9	0.27	4	1.55	19.41
56	RMP	Plant nutr'n-SG/PM/FM	4.5	13.0	29.0	0.08	0.07	32.1	12.4	0.70	3	1.54	19.49
57	LGM	Peanut mottle virus-GN	3.9	3.5	19.6	0.21	0.18	147.3	138.7	0.91	1	1.51	19.70

7 Using Multiple Objectives in Participatory Assessment of International Livestock Research: Lessons Learned

Patti Kristjanson, Thomas Randolph, Philip Thornton, Robin Reid and James Ryan*

Abstract

This chapter describes a systematic quantitative research priority assessment exercise undertaken with the involvement of roughly 100 researchers and partners of the International Livestock Research Institute (ILRI) in 1999 for an array of research alternatives, as part of the development of a new Institute strategy. It describes the criteria employed for the economic, environmental, poverty, geographic and capacity-related impacts of a wide range of research themes. Lessons learned in applying such a framework and the challenges related to evaluating each criterion are discussed. Subsequently, the influences of recommendations arising from the exercise are explored, so as to appraise contributions to the evolution of ILRI's research focus and approach during the 10-year period since the priority assessment exercise was completed.

Keywords: Priority setting, priority assessment, livestock research, poverty reduction, environmental impact scoring

Description of Approach

A comprehensive International Livestock Research Institute (ILRI)-wide initiative was undertaken in 1999/2000 to develop and apply a priority assessment framework for livestock research. The approach taken and framework developed are described briefly here. Readers interested in more details about the results are referred to Thornton *et al.*, 2000. In this chapter, we focus on the lessons learned, almost 10 years later, in investing in such a comprehensive and quantitative priority assessment exercise. The framework used is general

* The first portion of this chapter draws on Randolph T.F., Kristjanson, P.M., Omamo, S.W., Odero, A.N., Thornton, P.K., Reid, R.S., Robinson, T. and Ryan, J.G. 2001. A framework for priority setting in international livestock research. *Research Evaluation* 10(3): 142–159.

enough to deal with many species of livestock and is based on a set of criteria, appraised through combining a variety of methods. Innovative features of the framework include the use of a global poverty data set, a reasonably comprehensive environmental impact scoring scheme and Monte Carlo simulation to help assess the uncertainty associated with the various indices.

Within a systematic priority assessment framework, the agreed-upon objectives for research need to be translated first into transparent, quantitative decision-making criteria. In our case, the following five criteria were chosen to embody the multiple objectives of ILRI's research as expressed in its institutional mission statement and mandate:

- expected economic impact;
- contribution to poverty alleviation;
- environmental impact;
- internationality of the problem and solution;
- contribution to capacity building and research efficiency.

The Institute's potential options for the research agenda were represented by a set of research themes. These themes in general are concerned with research directly, but the framework is potentially flexible enough to address related activities, such as capacity building. To permit comparison, themes were assumed to conform to a simple model of the research process (Fig. 7.1). Each research theme was treated as if it were only one project covering a fixed number of years (*X* in Fig. 7.1.), with an estimated probability of success in achieving planned milestones and generating the Institute's intended research

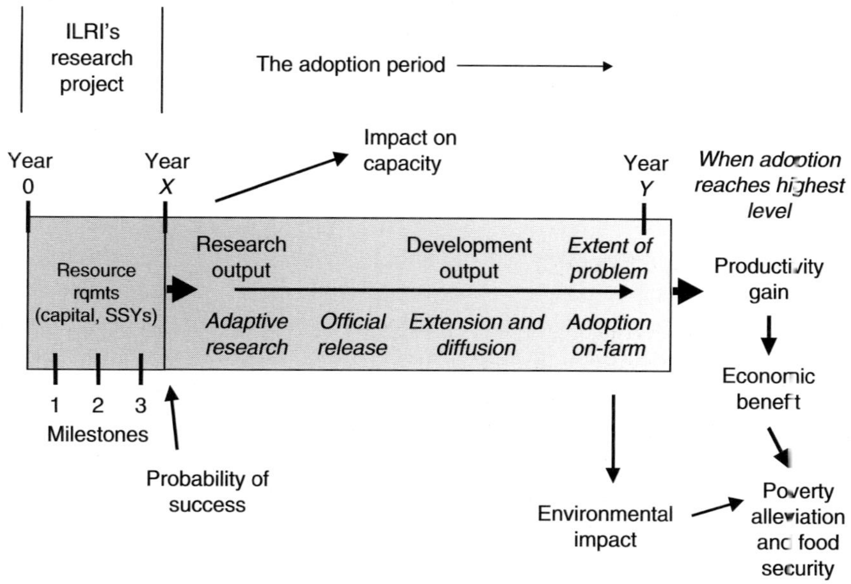

Fig. 7.1. Conceptual model of a research theme used in the priority assessment framework.

output. In cases where research was expected to generate outputs continuously over the life of the project, research and other outputs were assumed to be generated midway through the life of the project.

Resources required to achieve the objectives were measured in terms of scientific human resources (including operating costs) and new capital investments. Once the intended research output has been generated, there follows an envisaged process undertaken by other actors of further adaptive research, if needed, or a development of products customized to specific geographic regions, production systems or sets of end-users. This may entail a process of evaluation by National Agricultural Research Systems (NARS) or a commercial company before the product can officially be introduced in a given country. The product is then disseminated to end-users through formal extension channels or informal delivery systems such as non-governmental organizations or the private sector.

Adoption of the end product was assumed to begin immediately at the end of the research project and to follow the standard sigmoidal curve. The period between the end of the research project and attainment of the adoption ceiling constitutes the adoption lag. The estimate of the adoption ceiling was assumed to be made up of two elements: the extent of the problem being addressed by the research in terms of geographical coverage and production systems affected – which we term the relevance domain – and the portion of producers within that domain likely to adopt the end product.

Estimates for the various parameters defining each research theme were derived from a combination of scientists' expert opinions and available data. To facilitate estimation of key parameters such as probability of research success or eventual adoption rates, researchers were asked to use relative ranges of low, medium and high (based on existing knowledge or their best judgements), the exact values for which were decided later by a review committee of scientists. To permit quantifying the adoption of research outputs, the global livestock production systems classification of Seré and Steinfeld (1996) was used (see Randolph *et al.*, 2001 for a description of these 11 systems) to define adoption domains.

The simple model in Fig. 7.1 also describes the types of impact expected as a result of the research project. The principal pathway for impact is through improving productivity of smallholder livestock production systems. Adoption of the anticipated research products leads to productivity gains on farm, typically measured as increased output per animal of meat, eggs and milk. These gains then translate into economic benefits in terms of improved incomes, possibly through improved profitability (lower production costs), improved revenues (expanded production) and lower produce prices (of benefit to consumers) (Alston *et al.*, 1995).

Clearly, not all research themes affect on-farm productivity directly. Some research themes may focus on natural resource management and may indirectly affect incomes through better farm management (e.g. manure application) and not through increased livestock productivity *per se*. Other research themes focusing on improved decision making and policy research may be viewed as enhancing research efficiency, thereby accelerating the generation and transfer of research products more directly aimed at improving on-farm productivity. Unfortunately, little or no evidence is currently available as to the degree to which this acceleration actually generates incremental productivity gains at the farm level.

In addition to productivity gains leading to economic benefits, three other types of impact were associated with each research theme (Fig. 7.1). Contributions to capacity building may occur directly during the research project (training, information and networking) or be its primary output and, like policy research, improve research efficiency and create an accelerator effect. Second, adoption of the end product may have positive or negative effects on the environment, due either to the nature of the end product (e.g. technologies involving toxic chemical use) or to the implications that productivity gains and enhanced profitability may have for expanding livestock production. Third, depending on the proportion of poor among the beneficiaries of the research products, this environmental impact – because it affects the quality of the primary productive asset of many poor, namely land and water – when combined with the economic impact determines the ultimate impact on alleviating poverty and improving food security.

Criteria in the framework

The five quantitative decision-making criteria that formed the basis of ILRI's priority assessment exercise, and were applied to all areas of research, are described below.

Expected economic impact

One criterion for comparing the worthiness of different research themes is the economic impact that the research is expected to generate. In developing the economic impact indicator, we first identified and valued the potential incremental productivity gains attributable to the research outcome. Productivity gains and changing production costs are likely to change the amount of the commodity or commodities supplied. This market impact can be represented as a series of shifts in the supply curve and consequent adjustments in the equilibrium market price and quantity traded. These adjustments are captured by using an economic surplus model (see Alston *et al.*, 1995). The stream of estimated future benefits appropriately adjusted to account for the market effects and the time value of money was compared with the initial stream of research investments using a standard cost/benefit analysis.

For each research theme we then estimated the benefit/cost ratio (BCR), which essentially measures the value of the productivity gain generated per dollar invested in that research. In the framework, the BCR serves as the indicator for comparing the economic impact across research themes. The expected BCR for each theme is derived as the ratio of the discounted value stream of expected benefits to the discounted value stream of costs associated with the theme over a 50-year time frame. To be consistent with the other priority assessment criteria described below, the BCRs for all research themes were normalized to an index ranging from 0 for the lowest BCR to 1 for the highest. Following the suggestion of Alston *et al.* (1995), and agreeing with their justification, we used a discount rate of 5% for all analyses.

Poverty

Given the emphasis placed on poverty alleviation in the ILRI mission statement, a key indicator relates to the potential impacts that each research area may have in terms of affecting the number of poor in each region and production system. To capture potential impact with respect to both the extent and severity of poverty, we used an index computed as:

$$\textit{Poverty alleviation index} = \sum P_i x_i \Big/ \sum x_i,$$

where P_i is a measure of the severity of poverty in production system i weighted by x_i, the estimated number of poor people found in the production system.

The analysis was conducted at the level of 11 global livestock production systems to permit linking the relevant domains and adoption estimates for the research themes to the poverty alleviation impacts. These production systems often do not respect national boundaries, and so various manipulations were required to convert the available relevant data for constructing the index, which is often available only at the national or regional level, to the production system level.

The measure of the severity of poverty P is based on Gryseels *et al.* (1997), and is computed as $P_j = (1 - W_{ji}/Z_j)^a$, if $W_j < Z_j$, and as $P_j = 0$, if $W_j > Z_j$, where W is a welfare indicator, Z is a threshold income level per capita, a is an exponent that conditions the severity of the poverty index, and j is a country index.

The welfare indicator W is the average purchasing power parity (PPP) income (y), adjusted to reflect the degree of inequality in income distribution as measured by a Gini coefficient (G), and is computed as $W_j = (1 - G_j)y_j$. Thus, if the income distribution for the population within a country is highly skewed, G will be high, and the welfare indicator W for that country will be adjusted lower than for a country with equivalent income levels where incomes are distributed more equitably.

The indicator P thus provides a unitless index increasing from 0 to 1 as relative poverty becomes more severe. In this analysis, we estimated the index for the year 2010 to better match the period when impact is expected to occur. Country PPP income levels (y) reported in the UNDP Human Development Report (1994) were extrapolated to 2010 using per capita GDP growth rates from the data files of Gryseels *et al.* (1997). Gini coefficients are World Bank estimates from these data files. To calculate P, we used a threshold income level of US$6,000 per person, and set a to 2, following Gryseels *et al.* (1997). The estimated W and P values were converted from a country (j) to a production system (i) base for each of the six regions defined by Seré and Steinfeld (1996) using the average 1992–1994 human population data from Food and Agriculture Organization Statistical Database (FAOSTAT) (FAO, 1990–1998), and reallocated by livestock production system in Seré and Steinfeld's spreadsheets to derive human population by country and by production system. Country rural populations were allocated to the nine 'landed' systems in Seré and Steinfeld's classification based on the proportion of arable land in each agro-ecological zone (AEZ). Remaining population was allocated to the two 'landless' systems proportional to the total population in each AEZ.

To derive the poverty alleviation index for each research theme, an average weighted *P* value was calculated across the production systems of the relevant adoption domain with the number of poor in each system serving as the weights. Using the number of poor as the weighting factor effectively incorporates the extent of poverty into the index. The weights are based on estimates for 63 countries provided by Gryseels *et al*. (1997) for the total number of people below the 'poverty line' as defined by FAO and the proportion of rural versus urban poor.

The total number of poor was then allocated to the production systems in proportion to their human populations. These figures were further divided into rural and urban poor based on the country-level proportions of rural versus urban poor, and assuming that those in the 'landless' systems should be included with the urban poor.

Environment

The third criterion ILRI used to assess research was its potential environmental impact. Impacts on public health and genetic diversity of domestic plant and animal resources were also included here as 'externalities'. We assessed direct (immediate) impact of the intervention on four selected environmental properties: soil resources, water resources, greenhouse gas emissions and non-domesticated biodiversity. Under each of these properties, two to three subgroups were scored for impacts independently, and the average score was taken for each environmental property.

Indirect (longer-term) impacts of the intervention were estimated through two parameters: an index of the likelihood that the research would encourage extensification (expansion of agriculture), and the inherent fragility of the AEZ where the research would be applied. Public health impacts were restricted to the effect on zoonotic diseases.

DIRECT ENVIRONMENTAL IMPACTS. The impacts under each environmental property are summarized in Table 7.1. Under soil resources, we scored the impacts of the interventions on erosion (soil loss on site) and soil fertility (organic matter and nutrients). Water resources were divided into quality of water (levels of organic and inorganic nutrients, sediments, toxins) and quantity of water (water availability on site). Greenhouse gas emissions were separated into those that we consider to be most directly affected by livestock interventions: methane (CH_4), carbon dioxide (CO_2) and nitrous oxide (N_2O) emissions. We used two indices of non-domesticated biodiversity (which includes all taxa): species number as a proxy for genetic diversity and species composition. Biodiversity at the habitat or ecosystem level, estimated by species composition, was addressed further, though indirectly, through the index of the likelihood of extensification described below.

Table 7.1. Summary of environmental properties used to assess impacts.

Property	Weight	Description of impacts
Direct impacts		
Soil resources	0.125	Soil erosion
		Soil fertility
Water resources	0.125	Water quality
		Water quantity
Greenhouse gas emissions	0.125	Methane
		Carbon dioxide
		Nitrous oxide
Biodiversity	0.125	Species number
		Species composition
Indirect impacts		
Extensification	0.25	Likelihood of extensificaton
Habitat fragility	0.25	Fragility of AEZ

The scoring system used for direct environmental impacts includes five levels, ranging from strongly positive (+1) to strongly negative (−1). It is not possible to predict environmental impact for research that improves general research efficiency or dissemination (i.e. characterization, capacity building and information exchange); so for these types of research, we assigned the global mean of the environmental impact scores for all other research areas. Scoring was done by two ILRI ecologists.

For those research activities that may lead to an increase in animal production and thus animal numbers, we applied a general detrimental impact score to the direct impacts depending on the level of intensification of the system where the intervention is applied. For intensive systems, we gave a score of −0.5 and for extensive systems, −1.0. Increases in animal numbers were expected to have less impact in intensive systems because these systems are already highly used and further use will generate only marginal impacts. These scores were applied only to the environmental properties that may be directly affected by increased herd sizes: soil erosion, water quantity, methane emission, species number and species composition.

INDIRECT ENVIRONMENTAL IMPACT. Interventions that encourage extensification of agriculture were judged to have particularly great environmental impacts. These more indirect, system-level impacts were included because we anticipate, in many cases, that they will be greater and more important than the more immediate, local-level impacts. For example, the control of the livestock disease trypanosomosis may encourage and accelerate the expansion of agriculture on the agricultural frontier (Jordan, 1986).

The clearing of native forest and savannah has strong negative consequences for biodiversity, greenhouse gas emissions and soil and water resources. Fragility of the ecological region was included as an indirect impact although we are aware that the direct environmental impacts will be compounded in fragile habitats. Ecological regions are based on climatic, soil and topographic characteristics and on geographical location. Rainforest systems, for example, are particularly sensitive to clearing for livestock grazing, while arid grazing systems can be quite resistant to the impacts of livestock. Also, particular geographical locations are more sensitive to impacts than others. South American grazing systems, with a short history of evolution with grazing ungulates, are more sensitive to grazing impacts than African savannahs, where ungulates and grasses co-evolved over millennia together. Both the extensification and ecosystem fragility indices were scored as low (0.0), medium (−0.5) or high (−1.0).

PUBLIC HEALTH AND DOMESTICATED BIODIVERSITY IMPACTS. The public health index that was developed focuses on the impact of livestock interventions on the prevalence of zoonotic diseases. Control of animal trypanosomosis in Uganda, for example, may lead to direct control of human sleeping sickness as well. Domesticated biodiversity includes the impact of livestock interventions on the total store of domesticated breeds and species available for exploitation by humankind. As such, improvements in domesticated biodiversity can increase the number of species and varieties on earth, but this increase is likely a minor environmental impact compared with the potential loss of native species caused by increased livestock use around the world. This category was divided into species and breeds of animals and plants, specifically, livestock and fodder. Some ILRI research may attempt to improve and conserve the biodiversity of species and breeds of livestock and fodder. Some work may focus on attempting to conserve these species *in situ* through native habitat conservation, and these activities will also have direct and positive impacts on non-domesticated biodiversity. Only positive public health and domesticated biodiversity impacts were considered and these were scored between 0 and 1.

OVERALL ENVIRONMENTAL IMPACT SCORE. To produce the overall environmental impact score, the direct and indirect impacts were weighted equally. The overall 'externality' score was a weighted combination of the overall environmental score (0.90), the public health impact (0.05) and the domesticated biodiversity impact (0.05). Public health impact and domesticated biodiversity impact were given relatively low weights because it was anticipated that these impacts will be less widespread and important than the environmental impacts.

Internationality

Agroecosystems straddle national boundaries, as do major constraints to livestock development; consequently, so do research-induced opportunities to override these constraints. Considerable scope therefore exists to capture geographical spillovers in research output. The cross-national character, or 'internationality', of a given research theme was therefore considered to be a prominent feature in determining its priority ranking. As a measure of internationality, the Simpson index of diversity, I_k, is used in the framework: $I_k = \Sigma_m (S_{km}/100)^2$, where S_{km} is the share of economic returns to research theme k realized in a country or region or livestock system m. A variable $(1 - I_k)$ is defined such that a higher value indicated greater internationality. This variable thus gives greater priority to research activities that raise producer and consumer welfare in several parts of the world. A research activity thought to generate relatively small economic gains, but in several regions, thus had a higher internationality score than a theme that had a relatively large aggregate impact concentrated in one region or a few regions.

Capacity building and research efficiency

A research activity's contribution to capacity building and research collaboration with national agricultural research and development systems was identified as a key criterion in ILRI's priority assessment framework. In scoring this criterion, a disaggregated view was taken of each research activity by identifying if and how research activities and outputs were expected to have an impact on capacity building and collaborating to achieve research efficiency according to five sub-criteria:

1. Strengthened national human resources for research.
2. Strengthened national institutions for research.
3. Improved research tools adapted to national research needs.
4. Improved national human resources for development.
5. Improved national and local institutions for development.

A scoring scheme was used in which research activities and outputs that have a direct focus on any of these five sub-criteria were considered to have an 'important' impact, and it was given a value of 2. If activities and outputs have an indirect focus on any of these five sub-criteria, then its impact was considered to be 'incidental' and was given a value of 1. If activities and outputs did not focus on any of these five sub-criteria, then its impact was considered to be 'not applicable' and was given a value of 0. The maximum score that a research activity could attain for its impact on capacity building and research efficiency was therefore 10. As with the other criterion indices, the final scores were subsequently normalized across research themes to an index ranging from 0 to 1.

The composite index

Priority assessment based on the five criteria outlined above will inevitably entail trade-offs. One way to address these trade-offs is by representing them visually by a series of two-dimensional graphs where the likely trade-offs between pairs of criteria can be arrayed. Another complementary way is to take each normalized index and weight these to produce a single, integrated index for each theme. With an appropriate set of weights E_i on each theme k and criteria i, we arrived at a weighted average composite index CI_k, which combined normalized measures of each of the five criteria C_{ki} as follows:

$$CI_k = \Sigma_i C_{ki} E_i$$

Thus, with each component index normalized to range from 0 to 1, the resulting composite index for each research theme will be a number between 0 and 1. The normalization process and the weighting represent arbitrary scaling. However, if it is accepted that there are multiple objectives to be achieved in the conduct of publicly funded international agricultural research, and that often there are trade-offs among alternative research themes in their achievement, it is inescapable that some form of weighting must be used to assess thematic priorities. These can be either explicit or implicit. The composite index approach makes the process explicit.

This framework allows extensive sensitivity analysis to be undertaken, so that the robustness of the priority assessment can be gauged. In addition, Monte Carlo techniques can be employed on the quantitative components of the criteria in recognition of the fact that single-point estimates of variables such as the likely productivity gain from research or the probability of research success have substantial uncertainty attached to them. Probability distributions can then be imputed to such key variables, allowing confidence intervals for those variables to be generated through stochastic simulation. Such analysis can help in evaluating how robust a ranking of candidate research themes is with respect to the uncertainties involved.

Discussion

The extensive participatory process and the results are documented fully in Thornton *et al*. (2000). Here we briefly summarize some of the lessons learned in applying this framework, and with the benefit of hindsight, reflect upon the accuracy and usefulness of some of the predictions made.

In terms of the rankings of research themes across ILRI, the results generally matched scientist expectations. For instance, the livestock policy and capacity-strengthening research areas were ranked uniformly highly. This can be attributed to a large extent to their expected impact over a broad range of production systems, even assuming very conservative impacts on productivity. Themes for policy research were rewarded in particular for targeting poverty and generating positive environmental impacts. Research with an environmental focus also scored relatively highly. Environmental themes benefit not only from perceived positive environmental impact, but also from their ability to generate economic benefits and to target production systems with a high concentration of poor people. Themes in the livestock feeds and nutrition research area, on the other hand, were found across the full range of rankings. Those feed and nutrition themes that fell in the lower half of the ranking involved longer-term, higher-risk research.

The priority assessment results fed into ILRI's research planning over the following five to six years as one of the several inputs and considerations supporting decision making. For example, the results from the prioritization exercise contributed to a reduced emphasis on pastoral system issues in view of the fact that a relatively small share of the world's poor livestock keepers are found in this system, leading to a comparatively low-potential impact for research efforts (although this research area is now expanding in light of increased donor interest in environmental sustainability, niche markets and carbon sequestration). Some research areas were dropped (or 'turned over' to our partners), including prevention and control of trypanosomosis and breeding for improved feed utilization efficiency (both fell in the lower half of our overall ranking). Other potential research areas, even those that ranked highly in the analysis, including several of the environmental themes, were not pursued (e.g. reducing deforestation and reducing environmental costs of intensive livestock systems scored highly, but a management decision was made to leave this research to other, better-placed, institutions to pursue). Our policy and capacity-building research themes, with composite indices in the top half of the ranking, lost their programme status and were spread across several other research areas. This decision did not appear to be linked to the results of the prioritization exercise (in fact it was not obvious to any of the analysts what, if any decisions, were based on these results). One of the strategic directions that did come out of this activity, although perhaps in a latent way, was the need to understand more about where poor people are located, leading to seven years of involvement in capacity building with partners on poverty mapping.

In 2000, there were 26 rather loosely defined research themes evaluated in this exercise. By 2008, ILRI had reduced this number to 12 'Operating Projects', each typically made up of several donor-restricted/donor-funded projects, with much more coherent outputs defined. One lesson pertinent to the new use of medium-term plan (MTP) outputs at the institute level is that applying this approach is more difficult the further you 'scale up' from the individual theme level.

For institutions other than ILRI, of interest from a methodological standpoint is the performance of the indicators themselves and the robustness of the results in terms of sensitivity analysis and Monte Carlo simulation. We discuss these briefly below.

Performance of the indicators

Ideally, a quantitative assessment of research priorities provides information that permits decision makers to differentiate clearly among proposed themes in terms of their relative worthiness. Our results, however, showed many research themes having approximately equivalent composite index scores clustered together in a narrow range between 0.35 and 0.45. The clustering of results for the composite index was due to the counter-balancing effects of the five component indices and the nature of the distributions of the underlying component indices. In the case of the economic benefit index, the results are skewed heavily to the right, with substantial differentiation at the higher range of the scale, but with most of the lower values clustered under 0.05. This type of distribution permits identification of clear 'winners' but is less useful in distinguishing between the candidate themes at the lower end of the range, which is where resource constraints are most likely to require decisions to be made as to whether or not to undertake the research activities within a theme.

For the poverty impact and environmental impact indices, values were generally distributed smoothly across the full range, although they tended to cluster in the middle as a result of having to apply the same sets of assumptions to particular sets of research themes (e.g. in the policy area). The capacity-building index exhibited a step distribution, reflecting its rather simple structure of five component indicators, each with three levels. The internationality index distribution had the majority of themes clustered at the top end of the index range between 1.0 and 0.9, reflecting the initial screening that occurs when defining research themes appropriate for an international institute, ensuring that they address wide recommendation domains, and consequently making it difficult to distinguish significant differences in internationality impact between research areas.

In refining the priority assessment framework in the future, it may be beneficial to further develop the economic benefit and internationality indices, to improve their differentiating power. It is interesting to note that none of the correlation coefficients between the five component indices was significant at the 5% level. The lack of high positive or negative correlation coefficients, which would have had a reinforcing or balancing effect on the composite index, indicates that the current formulation of the indices does not contribute to any significant 'double-counting': the indices are indeed measuring different dimensions of impact. Our analysis indicated that trade-offs must be made between the various impact criteria in assembling a research portfolio. The composite index was useful, but the make-up of the individual research theme impacts also has to be considered.

The results of applying the framework to a set of candidate research activities for ILRI over a 10-year period indicated the importance of taking an explicit portfolio approach to livestock research. This process did not identify any 'wonder' research themes that scored highly in all aspects of the chosen criteria. While we do not have evidence to support the notion, we suspect that this would be a common finding, whatever the nature of the candidate themes in livestock research. Much more realistically, in our view, the results highlight the fact that research managers have to trade-off different objectives, and that this trade-off can only be made with a portfolio approach. Only by considering the totality of the research activities and their likely impacts together can the portfolio adequately address the goals of an institute such as ILRI. One area of possible future work on the framework would be to consider explicitly the variability of the various criteria that make up the composite index for each theme and to incorporate this into the framework explicitly, perhaps using approaches similar to those used in portfolio selection for stocks and shares.

One of the strengths of the framework outlined above is that it can be used to assign probability distributions to key uncertain or unknowable parameters (in our case, probability of success and productivity gain), and then Monte Carlo sampling of these can be carried out to produce not a single value of the BCR for all research themes but a probability distribution. In this way, some idea of the impact of the uncertainty on the analysis can be gleaned, and information concerning the uncertainty of the estimates of the indices for each research theme can point up important differences that may illuminate resource allocation decisions. On the other hand, for our analysis, assigning probability distributions did not significantly influence our composite index ranking, since they applied only to the BCR calculation and not to the other criteria

Various weaknesses in the framework can also be identified. One area concerns the reliability of the information related to expert input, such as in the identification of potential research activities. Perhaps more formal techniques that help to overcome bias could be developed. Another major constraint to the effective use of the framework was the availability and quality of data. Again, for reasons of general applicability, but also because of the nature of livestock, much of what might be considered the basic system's data for livestock-based production activities is highly aggregated and/or of doubtful quality. Refinement of the poverty and production system data is one example. Mention has been made above of the desirability of revisiting some of the component indices such as the environmental index, which could be developed further to include site-specific effects (some interventions having positive impacts in one place and negative impacts in another) and more extensive treatment of the public health impacts of proposed research themes.

Our framework, based on a relatively simple model of research and its impact, proved useful for summarizing some of the essential characteristics of potential research thrusts. However, as with any modelling framework, it could not address all of the complexities inherent in livestock (or indeed any) research. Some of the issues that were especially problematic included:

- identifying and valuing spillover effects, such as the value of scientific advances in veterinary medicine that might spillover into human health research;
- characterizing risks beyond the control of the research institute associated with adaptation and delivery from the time that the research ends until impact is achieved;
- defining roughly comparable research themes – the initial mix of themes included some that were very short-term, highly targeted projects and others that were longer-term, broader research thrusts.

In addition, the framework did not include any consideration of certain social issues, such as whether research outputs are socially appropriate for smallholders in the various recommendation domains, and how these outputs might affect women and children.

An interesting lesson, in reflecting on the value of this exercise, is that it can only be completed with the information available at the time, but there are many issues that can arise quite suddenly that change priorities, such as oil price shocks, use of feedstocks for biofuels and a global food crisis. Climate-change research was not on the ILRI agenda just 2 years ago, but is now substantial. Could these issues have been foreseen? Perhaps certain ones could have been, but others, such as changes in political salience, may be more difficult to forecast.

Conclusion

ILRI has made important progress in terms of identifying a strategic focus since this priority assessment exercise was undertaken. Though this progress cannot all be attributed to this one exercise, it certainly set the scene for its current strategy that focuses on poverty and a new geographic focus on South Asia and sub-Saharan Africa as areas of primary concern (ILRI, 2006). Since this exercise, more targeted, mostly qualitative, exercises have been carried out within each research theme to arrive at a few, clear aggregate outputs for each theme, linked carefully to project research activities with timelines and milestones towards achieving them. In other words, ILRI is now better able to link individual work plans with higher-level outputs and desired outcomes (with our wide range of partners), which allows it to better communicate its programme to the world in a fairly concise way. Were the numbers resulting from the quantitative priority assessment investment needed to get there? Probably not, but the process itself, and the conceptual framework used, was instrumental in getting all ILRI research teams to focus on where, what and how impacts from their work could be achieved. Now the institute is also finding other complementary processes, such as outcome mapping (see www.outcomemapping.ca), helpful for better understanding how different approaches within a project can best assist our partners achieve their desired outcomes and impacts.

References

Alston, J.M., Norton, G.W. and Pardey, P.G. (1995) *Science Under Scarcity: Principles and Practice for Agricultural Research and Priority Setting*. Cornell University Press Ithaca, New York.

FAO (1990–1998) *FAOSTAT-PC, FAOSTAT Statistics Database*. FAO, Rome, Italy.

Gryseels, G., Groenewold, J.P., Kassam, A. (1997) *TAC Database for Quantitative Analysis of CGIAR Priorities and Strategies*. TAC Secretariat, FAO, Rome, Italy.

ILRI (2006) *ILRI Annual Report 2005: Knowledge to Action: Tools for Livestock Development*. ILRI, Nairobi, Kenya.

Jordan, A.M. (1986) *Trypanosomiasis Control and African Rural Development*. Longman, London.

Randolph, T.F., Kristjanson, P.M., Omamo, S.W., Odero, A.N., Thornton, P.K., Reid, R.S., Robinson, T. and Ryan, J.G. (2001) A framework for priority setting in international livestock research. *Research Evaluation* 10(3), 142–159.

Seré, C. and Steinfeld, H. (1996) *World Livestock Systems: Current Status, Issues and Trends*. FAO Animal Production and Health Paper 127, FAO, Rome, Italy.

Thornton, P.K., Randolph, T.F., Kristjanson, P.M., Omamo, S.W., Odero, A.N. and Ryan, J.G. (2000) *Priority Assessment for the International Livestock Research Institute, 2000–2010*. Impact Assessment Series No. 6, ILRI, Nairobi, Kenya.

UNDP (1994) *UNDP Human Development Report*. Oxford University Press, Oxford

8 Priority Assessment for Rice Research in Sub-Saharan Africa

Aliou Diagne, Patrick Kormawa, Ousmane Youm, Shellemiah Keya and Simon N'cho

Abstract

Priority assessment is an integral part of research planning at the Africa Rice Center (WARDA), which involves a broad base of rice research and development stakeholders at different policy- and decision-making levels. Research priorities identified through a multi-stakeholder participatory process are reviewed, monitored and approved continuously by the WARDA Board of Trustees, the Science Council of the Consultative Group on International Agricultural Research (CGIAR) and the Council of Ministers. This chapter summarizes the methods and processes used in assessing research priorities at WARDA during the past 10 years. The essential elements of the priority assessment methodology were: (i) review the WARDA strategic plan (SP), its mission statement and goals to select *impact criteria* associated with the main developmental objectives towards which the medium-term plan (MTP) projects were intended to contribute; (ii) conduct a survey of representatives of major WARDA stakeholders to elicit relative (subjective) weights they attached to each selected impact criteria; (iii) develop a method and criteria for assessing and rating the potential and expected contribution of each research activity, output and programme/project to each selected impact criteria; and (iv) conduct a two-day priority assessment workshop involving WARDA management, scientists and representatives of major stakeholders to: (i) review and validate the methodology; (ii) conduct the assessment and ratings of the potential contributions of the research activities and outputs and the external factors that affect the conduct of research activities; and (iii) analyse and discuss the results of the priority assessment.

Keywords: WARDA, priority assessment, priority setting, rice, scoring

Introduction

Agricultural research organizations need an explicit and systematic process for setting research priorities effectively. This process is necessary to identify priority research themes and target populations, efficiently allocate scarce resources to priority research themes and enhance research relevance and impact on the livelihoods and well-being of the target population.

Priority assessment is an integral part of research planning at WARDA. Although WARDA is one of the 15 CGIAR Center its priority assessment process is unique. This is dictated by WARDA's partnership with 22 member states that demand services from the Center. The priority assessment process involves a broad base of rice researcher and stakeholder involvement at different policy- and decision-making levels. In addition to internally driven priority assessment exercises, priority assessment is conducted in the context of regularly held statutory meetings of the Council of Ministers (COM), the National Experts Committee (NEC), which is a technical advisory body of the COM whose members are Directors of the National Agricultural Research Institutes (NARIs) in the different member countries, the governing bodies of the operational networks (the West Africa Rice Research Network (ROCARIZ), the Inland Valley Consortium (IVC), East and Central Africa Rice research network (ECARRN) and the Africa Rice Initiative (ARI). Priority assessment is also informed through consultations and workshops that include the participation of non-governmental organizations (NGOs), development partners and subregional organizations such as the West and Central African Council for Research and Development (WECARD/CORAF), the Forum for Agricultural Research in Africa (FARA) and the CGIAR Science Council (CGIAR SC).

Research priorities identified through this multi-stakeholder participatory process are reviewed, monitored and approved continuously by the WARDA Board of Trustees, the CGIAR and the COM. These various channels and levels of assessing and validating priorities for rice research ensure that the priorities of National Agricultural Research and Extension Systems' (NARES) partners are aligned with those of WARDA and with the CGIAR broad vision of agricultural research in sub-Saharan Africa (SSA). Furthermore, the involvement of stakeholders in research and development planning ensures support for research and receptivity to, and sense of, ownership of research products, which are two key ingredients for the uptake and impact of research products.

This chapter summarizes the methods and processes used in assessing research priorities at WARDA during the past 10 years. In the second section, we describe briefly the methods and processes used during the 1990s to set priorities for rice research in West Africa and to produce the WARDA Main Term Plan (MTP) that guided WARDA research from 1995 to 2005. In the third section, we provide a detailed account of the process and methods for the more recent systematic priority assessment exercise at WARDA for the 2005–2007 MTP.

Priority Assessment for Regional Rice Research and the WARDA MTP in the 1990s[1]

Priority assessment for the WARDA MTP from 1994 to 2005 was implemented through a series of planning meetings (within and outside of WARDA),

[1] See WARDA (1993, 1997, 2001a,b) and CGIAR/TAC, 2001 for more details on priority setting at WARDA during the 1990s.

workshops and consultations with stakeholders. A three-step process of systematic data gathering, analysis and identification of research themes, programmes and projects was implemented during a 3-year period (1991–1993) to produce the 1994–1998 MTP. The first step consisted of country-level assessments of rice production and marketing constraints by National Agricultural Research Systems (NARS) and WARDA researchers using surveys and workshops. Data on the different rice ecologies and production systems existing in each WARDA member country and on the different production and marketing constraints encountered by farmers were collected from secondary sources and surveys of experts and key informants (researchers, extension and development workers). For each country, the relative importance of each rice ecology was assessed based on area, production (actual and potential) and population. The relative importance of each constraint found in the different rice ecologies was also assessed based on yield loss, scale and frequency. The results of this first step were a series of matrix tables by agro-ecology listing the major production and marketing constraints found in WARDA member countries with an indication of the relative importance of each constraint in each country in terms of high, medium and low (or using score in a scale of 1–5) (WARDA, 1993, 1997, 1999, 2001a,b).

The second step consisted of combining the country-level assessments with information generated from the different research projects that were being implemented within the 1990–1994 WARDA MTP to arrive at an aggregate assessment of the relative importance of each ecology in the context of identified constraints at the regional level. For example, in 1991, the priority assessment started with a planning meeting of the regional rice task forces. Scientists nominated to represent national programmes identified and ranked the biophysical production constraints within each major rice-growing ecology of their respective countries. These qualitative assessments provided a first estimate of production problems. A reduced set of the most significant constraints derived in this way formed the basis of developing research priorities for planning regional collaborative research (CGIAR/TAC, 2001, pp. 19–20). The priority ecologies and constraints on which regionally coordinated rice research would focus were then selected based on their relative importance and the probability of success from the research being undertaken in having a significant impact, in terms of the WARDA mission and goals, as stated in the 1990–2000 Center strategic plan (SP). Through the task forces, regional plans were developed for sharing tasks in key thematic areas. The assignment of task was based on an assessment of priority constraint and comparative advantage of each NARS and WARDA in conducting research to address those constraints. (CGIAR/TAC, 1993) The second step also involved an assessment of each countries' research capacities for each priority ecology/constraint, the assessment of the comparative advantages of WARDA or a NARS in taking the lead in research for each ecology/constraint and the constitution or reconfiguration of research groups called 'task forces' along disciplines and or ecologies. Each of these task forces focuses on a set of constraints/ecologies consistent with the disciplines making up the groups (breeding task force, Integrated Pest Management (IPM) task force, Natural Resource Management (NRM) task force, economics task force, mangrove rice task force, irrigated task force). It should be noted that there is a continuum among ecologies because common constraints occur within each of the major ecologies. For example, drought runs across all three ecologies, while iron toxicity is found only in lowland and irrigated ecologies, and soil acidity is found on the upland and hydromorphic phase of upland.

One of the main outcomes of the second step was the determination that the rice research conducted or coordinated by WARDA would focus on the three main ecologies: upland, lowland and irrigated. The research on the mangrove ecology, which was marginal and covered only six countries (Gambia, Guinea Conakry, Guinea Bissau, Nigeria, Sierra Leone and Senegal), was devolved in 1995 to the Rokupr rice research station in Sierra Leone. Special attention given to reviewing the results of past rice research in West Africa led to two important outcomes. WARDA produced *state-of-the-art* papers in several major disciplines/activity areas which helped to ensure that full advantage was taken of useful results, so as to avoid duplication of previous work. For instance, project proposals in the SP and the 1990–1994 MTP were critically examined and modified in the light of these reviews leading to more sharply focused projects in 1992–1994 and 1994–1998 MTPs (CGIAR/TAC, 1993, p. 27).

In the third step, WARDA scientists used the outcomes of the second step and worked in a series of group sessions to formulate the operational research projects for the 1994–1998 MTP. For those ecologies and production and marketing constraints identified in the second step as being regionally important and within the comparative advantage of WARDA, well-defined projects were formulated and each scientist proposed research activities that he or she could contribute for each project. After several iterations to ensure within- and cross-project coherence and complementarities, together with review by management, the research projects were submitted for the 1994–1998 MTP. Parallel to the development of the WARDA MTP, each task force selected its own priority research themes within the set of constraints/ecology identified for it. These priority research themes formed the basis for task force members' discussion and approval of resource allocation for the research proposals submitted by NARS scientists each year.[2]

The development of the 1998–2000 MTP took the 1994–1998 MTP as a starting point and updated the assessment of the relative importance of the production and marketing constraints in the different rice ecologies using the knowledge generated through the work of the task forces and the diagnostics and characterization research conducted by WARDA scientists. An assessment of the lessons learned during the implementation of the 1994–1998 MTP (both in terms of scientific achievements and the efficient conduct of research) was also conducted along with an assessment of the external research environment (policy changes, a broadened development agenda and a dwindling agricultural research funding base). This led to the phasing out of some projects, the consolidation of

[2] Each task force met every year to decide on priority research themes for the following year, allocate research funds (provided by WARDA) to NARS researchers on a competitive basis, assess the results of previous funding and monitor progress. Each task force was chaired by a NARS researcher with a WARDA researcher playing a facilitating role.

many small projects into fewer, larger and easier to manage projects and the expansion of the social science research programme. In particular, the research project on malaria and health was phased out and two new projects on *ex ante* and *ex post* impact analysis were created for the explicit purpose of guiding WARDA management in future priority setting and resource allocation decision making in a more transparent and scientific manner. WARDA's capacity to monitor and assess its impact was also strengthened by the appointment of an impact assessment economist. The formulation of the new set of projects for the 1998–2000 MTP and beyond was accompanied by a new programmatic structure.[3]

Priority Assessment for the WARDA 2003–2012 SP and 2005–2007 MTP

In 2000, WARDA management appointed a working group to develop a new SP for the period 2001–2010, which updated the previous SP and also considered recommendations from the just-completed External Programme and Management Review (EPMR) and developments within the CGIAR and the SSA region (WARDA, 2001c).

The working group took stock of the outcomes of the previous priority assessments which were being updated through the annual meetings of the task forces and through the research conducted by WARDA scientists. This information was complemented by a reassessment of the main lessons learned from the implementation of the 1990–2000 SP and a knowledge gap analysis. The lessons learned through the implementation of the 1990–2000 SP included the following:

- Postharvest processes can contribute significantly to improving rice productivity and profitability.
- Traditional African rice germplasm (including *Oryza* species) needs to be exploited efficiently (e.g. in the development of New Rices for Africa (NERICAs)).
- Farmer participation in technology development and transfer may significantly increase research impact.
- Decisions taken by farmers to adopt a new technology are conditioned by their resource status; therefore, there is a need to provide alternatives (or options) instead of rigid recommendations.
- Decisions taken along the production–consumption chain are conditioned to a certain degree by the policy environment.
- In addition to WARDA and traditional partners involved in the conventional production–consumption chain (i.e. NARS, extension agencies, rice processors, consumers), emerging rice stakeholders also include banks, bilateral agencies, ministries and local NGOs.

[3] Except for few modifications following the 2000 EPMR, the 1998–2000 MTP was guiding WARDA research up to 2004 because of delays in the development of the WARDA 10-year strategic plan (which was supposed to cover the period 2000–2010) and the civil war that erupted in 2002 in Côte d'Ivoire where WARDA had its headquarters.

The knowledge gap analysis was conducted by the working group through an iterative process consisting of brainstorming meetings, writing and reviewing background papers, and soliciting comments (both within and outside WARDA) on numerous drafts of the SP. The relative importance of the constraints and the potential impact of research on the constraints were assessed using a scoring method based on scientists' expert judgements. A review of the trends in domestic rice production, consumption and importation in SSA, and recent developments in the domestic policy environment and the international rice market, was also conducted. All member countries were consulted through the task force mechanism (ROCARIZ), as well as WARDA NEC and COM. The NEC had two meetings on this topic.[4] The Board also spent considerable time discussing the draft SP.

Through the knowledge gap analysis a number of new research questions emerged, which included: (i) How can the quality of locally produced rice be improved by better postharvest and processing? (ii) How can rice value chains be developed? (iii) How can the *glaberrima* germplasm in Africa be exploited in order to improve productivity and enhance diversity? (iv) How can improved water management practices be promoted for sustainable rice production? (v) What participatory approaches are effective in enhancing technology generation and adoption? (vi) Is the current average annual productivity gain of 1.8% sufficient to cope with increasing rice demand?

Development of the SP and MTP

The SP was developed through extensive internal and regional consultations and planning meetings, which took 3 years to complete and to arrive at a final SP document that was satisfactory to the WARDA Board of Trustees, the NEC and the COM (WARDA, 2004b).[5] That document, which ended up being the SP for the 2003–2012 period, incorporated the main elements of the SSA strategies of WECARD/CORAF (WECARD/CORAF, 1999), FARA (FARA, 2003a,b) and the CGIAR (CGIAR, 2000) as well as elements of the United Nations' Millennium Development Goals (United Nations, 2002) and those of the New Partnership for Africa's Development (NEPAD, 2003).

The 2003–2012 SP retains the strategic decision to focus on the three main rice-growing ecologies (upland, rain-fed lowland and irrigated). Two core research areas that cut across the three major rice ecologies were identified in the SP (WARDA, 2004b). The first research area is *development of components for integrated rice production systems*. Two major research issues that also cut across all three major rice ecologies were formulated within this core research area: (i) how existing genetic resources can be used to develop rice

[4] For ECARRN, a network created by WARDA and the Association for Strengthening Agricultural Research in Eastern and Central Africa (ASARECA), the International Service for National Agricultural Research (ISNAR) process of priority setting with an elaborate and well-established process was used. (CGIAR, SC, 2008, p. 69).

[5] The delay was due to a large extent to the civil war that erupted in Côte d'Ivoire in 2002.

varieties that best fit or better optimize farmers' existing production systems and that are acceptable to both producers and consumers; and (ii) how resource-use efficiency can be further increased for more productive, profitable and socio-economically viable rice production systems. The second research area was *analysis of rice policy and development options*, with two major research issues also included within this core research area: (i) identifying factors that affect the price of rice and rice market dynamics, and how these factors in turn affect the competitiveness of local rice production; and (ii) assessing impacts that technical, policy and institutional changes in the rice sector have on the livelihoods and well-being of the poor.

In terms of geographic coverage, the 2003–2012 SP expands to include the whole of SSA as the potential target area of WARDA's research programme,[6] but gives priority to work in countries where: (i) one or more of WARDA's research priorities are relevant; (ii) rice is an important crop; and (iii) the situation is likely to remain favourable to achieving substantial impact during the next decade. Apart from the traditional member countries in West and Central Africa, the Center's coverage was to be extended to Uganda, Ethiopia, Rwanda, Tanzania, Kenya, Madagascar, Gabon, Congo Brazzaville, Democratic Republic of Congo, Burundi and Mozambique.

After the SP was approved, another working group of WARDA scientists was appointed to develop the 2005–2007 MTP. The working group met and consulted over 3 months to formulate four projects within each of the two core research areas identified in the 2003–2012 SP. The resulting eight projects addressed all the four research issues described above while taking into account the CGIAR systems priorities and its new guidelines for formulating MTP projects and outputs (size, scope and format) as well as relevant Challenge Programs (WARDA, 2004a). Three of the projects within the first core research area are ecology-focused and formulated to address the production constraints of the three major ecologies (upland, lowland and irrigated) and one is focused on drought, which was identified as a major constraint with significant impacts on poverty across both the upland and lowland ecologies. The four projects in the second core research area tackle research questions that cut across all ecologies and are related to the paucity of uptake of many of the improved agricultural technologies, the institutional environment, policy and market dynamics, postharvest processes, impact assessment and the mitigation of the effects of man-made and natural disasters on agricultural production.

Priority Assessment for the 2005–2007 MTP: Methodology and Results

Once the MTP projects, outputs and activities were formulated, resources had to be allocated to the different projects. For that purpose, an internal

[6] The 4th WARDA EPMR endorsed that the Center could cautiously expand its activities to East and Southern Africa based on availability of extra resources.

priority assessment task force was commissioned with the mandate to develop a transparent priority assessment methodology and approach for prioritizing the research projects, outputs and activities and for allocating resources among them. The task force reviewed the priority assessment methods and approaches used in the CGIAR and other non-CGIAR Center, subregional organizations (SROs) and the NARS. The task force then developed its own method, borrowing and adapting elements from the different methods reviewed. The essential elements of the methodology were the following:

- Review of the WARDA SP, its mission statement and goals to select *impact criteria* associated with the main developmental objectives towards which the MTP projects were supposed to contribute.
- Conduct of a survey of representatives of major WARDA stakeholders to elicit *relative (subjective) weights* they attached to each selected impact criterion.
- Develop a method and criteria for assessing and *rating the potential and expected contribution of each research activity, output and project* to each selected impact criteria.
- Conduct a *2-day priority assessment workshop* involving WARDA management, scientists and representatives of major stakeholders to: (i) review and validate the methodology; (ii) conduct the assessment and ratings of the potential contributions of the research activities and outputs and the external factors that affect the conduct of research activities; and (iii) analyse and discuss the results of the priority assessment.

The impact criteria

The task force developed a set of six impact criteria based on WARDA's mission statement and strategic goals as laid down in the 2003–2012 SP. Each criterion was defined and keywords provided to ensure that stakeholders had a similar understanding of the terms used. The six impact criteria and their associated keywords are given below:

1. *Contribution to poverty alleviation*: Income, employment for the poor, market access, education, gender, equity.
2. *Food and nutrition security and good health*: Increase yield, nutrition, enhanced micronutrient value, health, mitigating effects of HIV/AIDS malaria and nutrition-related diseases, disaster and conflict recovery, emergency seeds.
3. *Partnership/capacity building*: Training for partners, respond to end-user needs (NARES, scientists, farmers), stakeholder platform, networking, information exchange, participatory approach, stakeholder ownership, policy dialogue, advocacy.
4. *Increasing productivity and profitability*: Economic growth, food availability, foreign exchange savings, improved market access, improved market efficiency, increased resource-use efficiency.

5. *Sustainability of farming environment*: Maintain resource quality, soil quality recovered, reduce pests and diseases, protect groundwater and surface water quality, increase biodiversity, mitigate global warming, reduce deforestation, reduce human health hazards caused by input use.
6. *International public goods*: Publications, methodologies, models, varieties, improved production technologies and processes, improved crop management practices.

The impact criteria were included in a one-page questionnaire (Table 8.1) and sent to representatives of a wide range of WARDA stakeholder constituencies that included WARDA board members, NEC members and representatives of SROs and producer groups. The respondents were asked to assign a relative weight from 0 to 100 to each impact criterion based on their own assessment of the relative contribution of that impact criterion to achieving WARDA's mission and strategic goal and to achieving the millennium development goals. Assigned weights had to sum to 100.

Table 8.1. Questionnaire for eliciting stakeholders' relative weights for the six impact criteria.

Impact indicator	Weight on impact indicator (0–100%)
Poverty Alleviation: *Income, Employment for the poor, Market Access, Education, Gender, Equity*	
Food and nutrition security and good health: *increase yield, nutrition, enhanced micronutrient value, health, mitigating effects of HIV/AIDS and malaria, disaster and conflict recovery, emergency seeds*	
Partnership/capacity building: *training for partners, respond to end-user needs (NARES, scientists, farmers), stakeholder platform, networking, information exchange, participatory approach, stakeholder ownership, policy dialogue, advocacy*	
Increasing productivity and profitability: *economic growth, food availability, foreign exchange savings, improved market access, improved market efficiency, increased resource-use efficiency*	
Sustainability of farming environment: *maintain resource quality, soil quality recovered, reduces pests and diseases, protects groundwater and surface water quality, increased biodiversity, mitigates global warming, reduces deforestation, reduces human health hazards caused by input use*	
International public goods: *publications, methodologies, models, varieties, improved production technologies and processes, improved crop management practices*	
Total contribution to impact indicators (weighted)	100%

Assessing and rating the contribution of outputs to the impact criteria

After selecting the impact criteria, the next step was to devise a method to assess the potential contribution of each MTP project output to the impact criteria. The following three criteria with their associated keywords were selected by the task force[7]:

1. *Potential contribution of output if achieved and adopted/used*: Yield, labour-saving (more time for non-farm income, child education); input-use efficiency (less pesticide and fertilizer use), chemical-use efficiency; improved soil quality indicators (improved soil fertility, less erosion, less salinity, less acidity); low incidence of human health hazards from use of fewer chemical inputs; better functioning of input and output markets, more income, more food, more employment.
2. *Potential adoption/usage rate if output is disseminated*: Proportion of output users/adopters if they have access to output; proportion of poor served by the output; possibility of scaling up/out; cost of output to adopters; output requires high input use.
3. *Proportion of population with access to output*: Will the output reach end-users or remain on the shelf; potential interest of partners to disseminate output; cost of dissemination; political/donor support to disseminate output.

The first criterion assesses the potential impact of an output. It is in a sense a measure of its intrinsic value to a user. The second criterion assesses the extent to which an output will be used by the target population and it is in a sense a measure of the extent of the demand for the output and its value to society. The third criterion is a measure of the efforts that will be made to make the output available to its intended users. The three criteria are measures of aspects that are necessary for an output to have any impact. The aspects measured by the first criterion are directly related to research efforts while the aspects measured by the two other criteria are to a large extent dependent on factors outside the control of research organizations. The assessment was done by the participants of the priority assessment workshop (individually) for each output using a scoring method that varied from 1 to 5.[8] Subsequently, two questionnaires containing all projects outputs (Tables 8.2 and 8.3) were designed and distributed to all the participants in the workshop.

Assessing and rating the contribution of research activities to outputs

Each output is produced through the conduct of one or more research activities. Hence, it is necessary to have a method for assessing the effective contribution of each research activity to an output and the likelihood of achieving the

[7] Tables 8.2 and 8.3 below show the relation between these three criteria and the six impact criteria.
[8] 1 = very low, 2 = low, 3 = medium, 4 = high, 5 = very high.

Table 8.2. Questionnaire for assessing the potential contribution of project outputs to each impact criterion.

	Potential contribution of output if adopted/used by the target population			
Impact criteria	Output ×.1	Output ×.2	Output ×.3	Output ×.*n*
Poverty alleviation				
Food and nutrition security and good health				
Partnership/capacity building				
Increasing productivity and profitability				
Sustainability of farming environment				
International public goods				

Table 8.3. Questionnaire for assessing the extent of potential dissemination and adoption/usage of project outputs.

	Outputs			
	Output ×.1	Output ×.2	Output ×.3	Output ×.*n*
Potential adoption/usage rate of output if disseminated				
Proportion of population who will have access to output within next 25 years				

output in a cost-effective way. The following three criteria with their associated keywords were used to link the conduct of a research activity to the achievement of a project output:

1. *Percentage potential contribution in achieving output if successfully completed (scale 0–100 with total across activities adding up to 100 for each output)*: Is activity relevant to the achievement of output? Could the output be achieved without conducting this activity?
2. *Probability of successfully completing each activity if resources are available*: Is there availability of resources to complete the activity within the time frame (be forward-looking); financial (donor/core/restricted, donor interest, timely availability of funds); human (skilled/competent/adequate).
3. *Comparative advantage of WARDA in conducting the activity*: If the activity is not conducted by WARDA, will another CGIAR Center conduct it? If WARDA conducts it, will it use fewer resources than other Center would have used? Is WARDA better placed within the regional context to carry out the activity? Capital equipment and supplies (lab equipment and supplies, computers, software) are available.

The first criterion measures the relative intrinsic scientific value of the research activity, independent of external factors that are outside the control of

Table 8.4. Questionnaire for assessing the contribution of research activities to the likelihood of achieving each output.

	Outputs				
	Output ×.1			Output ×.2	
	Activity ×.1.1	Activity ×.1.2	Activity ×.1.*n*	Activity ×.2.1	Activity × 2.*n*
Percentage potential contribution in achieving output if successfully completed: (*scale 0–100 with total across activities not exceeding 100 for each output*)					
Probability of successfully completing each activity if resources are available: (*rating scale: 1 = very low, 2 = low, 3 = medium, 4 = high, 5 = very high*)					
Availability of resources to complete activity within time frame (be forward-looking): (*rating scale: 1 = very low, 2 = low, 3 = medium, 4 = high, 5 = very high*)					
Comparative advantage of WARDA in conducting activity: (*rating scale: 1 = very low, 2 = low, 3 = medium, 4 = high, 5 = very high*)					

researchers. The second criterion reflects the fact that there is no 100% guarantee that a solution to a research problem will be found within a reasonable time frame even if all the necessary human and financial resources are used in the research activity. The last criterion is an assessment of the extent to which the activity may be redundant for achieving the output or may be conducted more efficiently by another research institution. Similarly to the outputs, the assessment and rating of the activities were individually completed by the participants in the priority assessment workshop using the same scoring method (scores varied from 1 to 5) and the questionnaire shown in Table 8.4.

The two-day priority assessment workshop

A two-day priority assessment workshop was organized to review and validate the methodology developed by the task force and stage the priority assessment exercise. Participants in the workshop included WARDA scientists and management, representatives of WECARD/CORAF and NARS collaborators involved in the networks (ROCARIZ, ARI and IVC). The SP and the eight projects were presented first in a plenary session. Then, the participants were divided into working groups of multidisciplinary teams that critically reviewed each project, project outputs and activities against the sets of criteria. Before members of a working group started rating a project output and its activities, a WARDA scientist knowledgeable of the output presented it to the group and answered questions. The group would then discuss briefly the outputs and activities with respect to the criteria before each member was left to fill individually his or her two questionnaires for that output and its activities as described above. Each workshop participant also completed the impact criteria questionnaire which was sent to representatives of WARDA's major stakeholders prior to the workshop. It took a day and a half to complete the assessment of all the project outputs and activities. The questionnaires were collected one by one on completion and the data were immediately entered into spreadsheets by a team of WARDA research assistants. Statistical Package for the Social Sciences (SPSS) programmes for analysing the data and presenting the results were written at the same time these activities were taking place. This concurrent processing of the priority assessment data allowed the results to be presented and discussed by participants before the end of the second day of the priority assessment workshop.

Calculation of the potential and expected contributions of outputs and projects to each impact indicator and to total impact

In this section, we describe how the potential and expected contributions of each output and each project to each impact criterion and to the total impact were calculated from the data obtained from WARDA's stakeholders and the participants of the priority assessment workshop using the three questionnaires described above. The calculations were completed in three steps.

In the first step, the assessment of the potential and expected contributions of each output to *each impact criterion* and the likelihood of achieving each output were calculated at the individual respondent level (using the data from the priority assessment workshop participants – Tables 8.2, 8.3 and 8.4). The calculation of these individual respondents' assessments of the potential and expected contributions is explained in more detail below.

For each respondent, the potential contribution of each output to each impact criterion was obtained directly from the entries of Tables 8.2 and 8.3. The expected contribution to each impact criterion was obtained by multiplying the corresponding potential contribution in Table 8.2 by: (i) the potential adoption/usage rate of the output if disseminated (the corresponding entry in the first row of Table 8.3); (ii) the proportion of the target population who will have access to the output (the corresponding entry in the second row of Table 8.3); and (iii) the likelihood of achieving the output efficiently within the time frame.[9]

The likelihood of achieving an output effectively within the time frame was obtained from Table 8.4 in two steps. First, for each activity the following fac-

[9] The expected contribution of each output to each impact indicator is re-scaled so that its value is between 1 and 5. For that purpose, a linear transformation in the form of $y = ax + b$ is used; where $a = (1 - (1 - 1/25)/(5 - 1/25))$ and $b = (1 - 1/25)/(1 - (1/25)^{\cdot}(1/5))$.

tors were multiplied together: (i) the contributions of the activity in achieving the output (the corresponding column entry in the first row of Table 8.4); (ii) the probability scores of successfully completing the activity (the corresponding column entry in the second row of Table 8.4); (iii) the score on the availability of resources to complete the activity within time frame (the corresponding column entries in the third row of Table 8.4); and (iv) the score on the comparative advantage of WARDA in conducting the activity (the corresponding column entry in the last row of Table 8.4).[10] These products were added across all the activities making up that output (i.e. a summation across the columns for each output in Table 8.4).

In the second step of the calculation, the assessment of each participant in the priority assessment workshop of the potential contribution of each output to *total impact* was calculated by taking the summation of the respondent's assessment of the output's potential (expected) contributions to the six impact criteria as calculated in the first step. Both weighted (using the weights obtained from Table 8.1) and unweighted total impacts were calculated. The weighted total impact was obtained by weighting each respondent's assessment of an output's contribution to each impact criterion (potential and expected) by the mean relative weight that WARDA stakeholders attached to that impact criterion.[11] The mean relative weight of each impact criterion was obtained from the data collected with the stakeholders' questionnaire (Table 8.1) by taking the average score for that criterion across all respondents to the stakeholders' survey.[12]

In the third step, the series of individual respondents' assessments of the potential and expected contributions were aggregated across all respondents to calculate the means of potential and expected contributions of each output to each impact criterion and to total impact (weighted and unweighted) At this step, the mean contributions were also calculated for each MTP project (by summing each individual respondent's assessment across the project's outputs and then taking the mean across all respondents). Various features of the distribution of the individual respondent-level assessments were also computed or visualized through graphs (by output and by project).

Results of the prioritization: presentation and discussions

The results of the analysis of the data collected from the four questionnaires are presented in Tables 8.5–8.9.[13] The mean relative weights that WARDA

[10] The ordinal scores were converted to probability values ranging from 0.2 to 1 with 1 = very low = 0.2, 2 = low = 0.4, 3 = medium = 0.6, 5 = very high = 1.

[11] The unweighted total impact gives equal weight to each impact criterion.

[12] The number of respondents to the stakeholders' questionnaire is normally greater than those for the other three questionnaires (Table 8.2, 8.3 and 8.4).

[13] The return rate for the questionnaire sent to outside stakeholders (Table 8.1) was very low (less than 5%). The low return rate was probably due to the fact that it was sent just 2 days prior to the priority assessment workshop.

Table 8.5. Stakeholders' relative weights for the six impact criteria.

Impact	*N*	Mean*	SE (mean)	Median	Standard deviation
Poverty	26	18.65	1.58	20.00	8.07
Food security	26	18.65	1.06	20.00	5.40
Partnership	26	15.77	1.26	15.00	6.43
Productivity	26	18.85	1.83	20.00	9.31
Sustainability	26	18.08	1.27	17.50	6.49
Public good	26	10.00	0.78	10.00	4.00
Total	156	16.67	0.60	15.00	7.43

*The Bonferroni multiple-comparison test shows that the mean relative weights of the impact criteria are all not significantly statistically different (p-value > 0.10) except for the public good impact criterion which is significantly lower to all of the others (p-value < 0.01).

Table 8.6. Potential contributions of outputs to impact criteria.

Impact	*N*	Mean*	SE (mean)	Median	Standard deviation	Min	Max
Poverty	886	3.18	0.03	3.00	0.96	1	5
Food security	886	3.27	0.03	3.00	0.99	1	5
Partnership	886	3.35	0.03	3.00	1.02	1	5
Productivity	886	3.28	0.03	3.00	1.00	1	5
Sustainability	886	2.98	0.04	3.00	1.11	1	5
Public good	886	2.93	0.04	3.00	1.11	1	5
Total	5316	3.16	0.01	3.00	1.04	1	5

*The Bonferroni multiple-comparison test shows that the mean potential contribution for partnership is not significantly statistically different from that of productivity and that of food security (p-value > 0.10), while being significantly statistically higher than that of poverty (p-value = 0.10).

Table 8.7. Correlations matrix for the potential contributions of outputs to the impact criteria (*P*-value in parenthesis).

	Poverty	Food security	Partnership	Productivity	Sustainability	Public good
Poverty	1					
Food security	0. 561 (0.00)	1				
Partnership	−0.0983 (0.053)	−0.1465 (0.0002)	1			
Productivity	0.3770 (0.00)	0.4628 (0.00)	−0.1929 (0.00)	1		
Sustainability	0.2499 (0.00)	0.3246 (0.00)	−0.1377 (0.0006)	0.3869 (0.00)	1	
Public good	0.0911 (0.1032)	0.0761 (0.3599)	0.0341 (1.00)	0.1318 (0.0013)	0.1926 (0.00)	1

Table 8.8. Likelihood of achieving outputs, access to outputs and adoption of outputs by target population.

	N	Mean	SE (mean)	Median	Standard deviation	Min	Max
Likelihood of achieving the output	4968	0.40	0.00	0.38	0.18	0.00	1
Likelihood of access to the output by the target population	5310	0.61	0.00	0.60	0.22	0.20	1
Likelihood of adoption of outputs by the target population	5304	0.68	0.00	0.60	0.17	0.20	1

Table 8.9. Contribution of research activities to the likelihood of achieving each output.

Variables	*N*	Mean	SE (mean)	Median	Standard deviation	Min	Max
Relative contribution of activities	3027	0.2929	0.0029	0.25	0.1586	0	0.7
Likelihood of completing activities	3157	0.7510	0.0032	0.80	0.1826	0	1
Availability of resources to complete activities	3092	0.6530	0.0034	0.60	0.1905	0	1
Comparative advantage of WARDA in conducting activity	3143	0.7409	0.0033	0.80	0.1861	0	1

stakeholders attached to the six impact criteria are presented in Table 8.5. From the Table, we can conclude that on average WARDA stakeholders attached relatively lower weights to international public good (10) and partnership (15.77) compared to the other impact criteria (poverty, food security, productivity and sustainability) which all have mean relative weights of about 18.[14] On the other hand, we can see from Table 8.6 that the potential contributions of the outputs are greatest for the partnership impact criterion with a mean score of 3.35 (out of a maximum of 5) compared to 3.28 for productivity, 3.27 for food security, 3.18 for poverty, 2.98 for sustainability and 2.93 for international public good. The mean potential contribution of all outputs to total impact is 3.16. The Bonferroni multiple-comparison test shows that the mean potential contribution for partnership is not significantly statistically different from that of productivity and that of food security (p-value > 0.10), while being significantly statistically higher than that of poverty (p-value = 0.10).[15] The

[14] The mean relative weights of the impact criteria are all not significantly statistically different except for the international public good impact criterion which is significantly lower to all of the others.

[15] The mean potential contributions to the poverty, food security and productivity criteria are not statistically significantly different (P-value > 0.10).

mean potential contribution to each one of the four impact criteria (partnership, productivity, food security and poverty) is statistically significantly higher than that of sustainability and that of international public good. The matrix of correlations of the potential contribution of the outputs to the six impact criteria shows that the potential contribution to poverty and that to food security are strongly positively correlated with correlation coefficient above 0.5 and statistically different from zero (p-value < 0.01). The potential contribution to poverty is also significantly positively correlated to productivity and to sustainability (p-value < 0.01). The positive correlations of the potential contributions to poverty, food security and productivity should be expected because in principle both productivity and food security contribute to poverty reduction. On the other hand, the potential contributions to these four impact criteria (poverty, food security, productivity and sustainability) are significantly negatively correlated with that of partnership. This indicates some trade-off between spending efforts in building partnerships and working directly to achieve welfare impacts (as measured by the four impact criteria). Note that the international public good impact criterion is only significantly (positively) correlated with the productivity and sustainability criteria.

Table 8.8 shows that the mean likelihood of achieving the outputs is 40% while the mean likelihoods of accessibility and adoption of the outputs by the target populations are, respectively, 61% and 68%. The mean likelihood of achieving the outputs is relatively low because it is the weighted average across research activities (with the weights being the relative contribution of activities to achieving the outputs) of the product of three probabilities (see Table 8.4): (i) the likelihood of completing activities; (ii) the availability of resources to complete activities; and (iii) the comparative advantage of WARDA in conducting the activities. Table 8.9 shows the means for the three probabilities are relatively high. Indeed, the mean probability of a research activity being successfully completed when resources are available is 75% with an average chance of 65% that the resources to conduct the activity will be available. On average, the relative comparative advantage of WARDA to conduct a research activity is 74%. With respect to the contribution of research activities to the likelihood of achieving the outputs, Table 8.9 indicates that on average the relative potential contribution of a research activity to achieving an output is 30%.

The results were presented in a final plenary session. But, due to lack of time, the disaggregated results for the different outputs were not discussed in the workshop. Instead, the results were aggregated and presented at the project level. The project-level results are summarized in Figs 8.1 and 8.2. The results indicate that, in general, each project ranked medium to high when all impact indicators are taken together.[16] When external factors were not considered (or complete adoption is assumed), the average weighted *potential* contribution of all projects to all indicators except 'public or

[16] The presentation of the results at the aggregated project level tends to make the approach less able to discriminate among alternatives. The disaggregated output-level results show much more contrast in the rankings of the outputs.

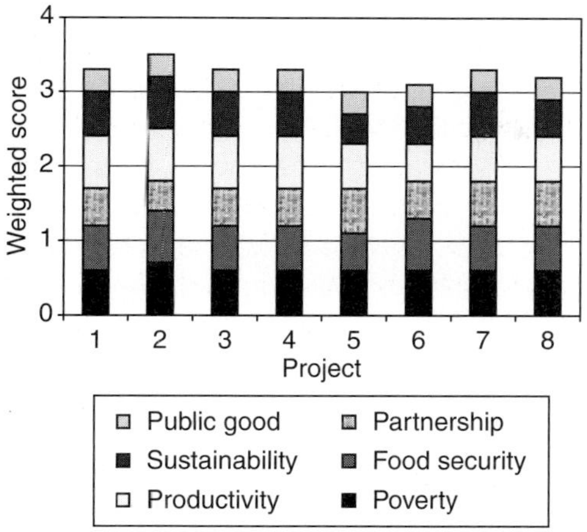

Fig. 8.1. Average weighted potential contribution of project outputs to each impact criterion and to total impact, assuming that complete adoption occurs.

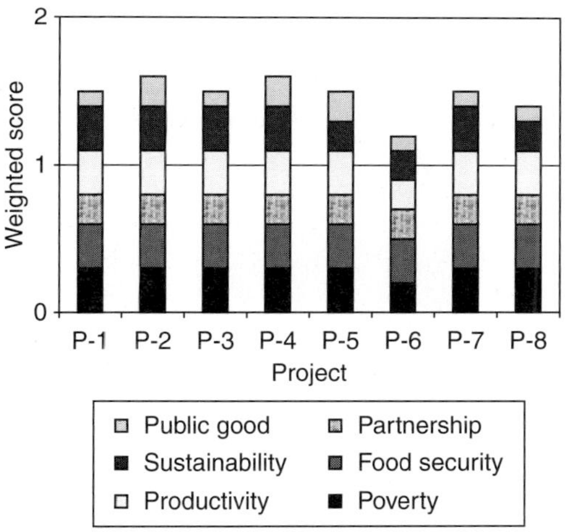

Fig. 8.2. Average weighted expected contribution of project outputs to each impact criterion and to total impact, considering the probability of research success, dissemination and adoption.

international good' fell between high and very high (Fig. 8.1). However, no one project ranked high for all criteria.

Figure 8.2 presents the average weighted, *expected* contribution of project outputs to each impact indicator and to the total impact when external factors affecting the probability of an output reaching and being used by an adopter are taken into consideration. While Fig. 8.1 considers only research-related factors, Fig. 8.2 considers external factors when ranking the outputs. The figures indicate that contributions to the different indicators scored from low to medium, except for partnerships and public or international good, which were from low to very low. These results were expected because the external factors that affect the achievability of research outputs are beyond the researcher's control. While this may appear simplistic, it highlights the fact that if external factors are controlled, the intrinsic contribution of the MTP projects will range from high to very high. However, factoring in external factors significantly reduces the potential contribution of the projects.

The relative rankings of the eight MTP projects with respect to each impact criterion and to all impact criteria were discussed by participants. In general, the discussion focused more on the complementarities among the projects in achieving the different goals as measured by the six impact criteria than on singling out specific projects for their performance. The output-disaggregated results were subsequently studied by the WARDA MTP working group and management after the workshop to refine the output contents of the MTP projects and to decide on the relative sizes of the eight projects in terms of budget and human resources. The eight projects were nested in two separate programmes: Integrated Production Systems and Rice Policy and Development. The former was assigned 60% of resources and the latter was assigned 40% in the 2003–2012 SP. As a result of the priority assessment exercise, the research aspects of the networks were placed under one project on partnerships. It was a responsibility of the management to ensure there is a resource allocation to the MTP projects and outputs consistent with the results of the priority assessment exercise. As unrestricted funding was not immediately available, resources had to be identified to kick-start the projects based on the priority allocated to each project.

Conclusion

Priority assessment is an integral part of research planning at WARDA and it is conducted systematically every five years through an internally driven process that follows a well-defined methodology to develop the SP and MTP and prioritize research. These reviews allow WARDA to adapt its priorities to new challenges and constraints.

Priority setting is a key function in any research institution's performance and growth, and provides an effective means of identifying areas of focus that will ensure impact. The methodology and approaches should be clear and simple enough to be repeatable in wider domains, rigorous and systematic. Multi-stakeholder involvement is a healthy part of the process and it validates the priorities and methodology and creates a greater sense of ownership. The priority assessment process paves the way for better and stronger horizontal and verticals links among partners, and is a way of shifting from 'business as usual' in light of needs for greater accountability. It provides a good platform for developing MTP and SP. The substantial investment in priority assessment in 2004 was essential in order to translate the SP 2003–2012 into reality and to align with new development in the new CGIAR vision, FARA and Comprehensive Africa Agriculture Development Programme (CAADP). Therefore, the subsequent MTPs only required minor annual revision.

References

CGIAR Science Council (2008) *Report of the 5th External Program and Management Review of the Africa Rice Center (WARDA)*. CGIAR Science Council Secretariat, Rome, Italy.

CGIAR/TAC and CGIAR Secretariat (2001) *Report of the Fourth External Program and Management Review of the West Africa Rice Development Association (WARDA)*. TAC Secretariat, FAO, Rome, Italy. 79 + Appendix.

CGIAR/TAC and CGIAR Secretariat (1993) *Report of the Third External Program and Management Review of the West Africa Rice Development Association (WARDA)*. TAC Secretariat, FAO, Rome, Italy.

CGIAR (2000) *Strategy for the Consultative Group on International Agricultural Research (CGIAR) in sub-Saharan Africa*, 18.

FARA (2003a) Sub-Saharan Africa challenge programme proposal. Building sustainable livelihoods through integrated agricultural research for development. *Securing the Future for Africa's Children* 1, 129.

FARA (2003b) *FARA's Strategy for Agricultural Research for Development in Africa, 2002–2012*. FARA, Accra, Ghana, 26.

NEPAD (2003) *Comprehensive Africa Agriculture Development Programme (CAADP)*. NEPAD Secretariat, Midrand, South Africa, 102.

United Nations (2002) *Implementing the Millennium Declaration: The Millennium Development Goals and the United Nations Role*. Fact Sheet. UN Department of Public Information, New York, 2.

WARDA (1993) *Medium Term Plan 1994–1998 for Presentation to TAC on the 25th of March 1993*. WARDA, Côte d'Ivoire, 122.

WARDA (1997) *WARDA Medium Term Plan 1998–2000 for Presentation to the Mid-term Meeting of the Consultative Group on International Agriculture Research, Cairo, Egypt 26–30 May 1997*. WARDA, Côte d'Ivoire. 81.

WARDA (1999) *Program Priorities and Strategies 1999–2005, Preliminary Draft*. WARDA, Côte d'Ivoire, 141.

WARDA (2001a) Potential for Green Revolution in Rice in West and Central Africa. Report of Second Biennial WARDA/National Committee Meeting, 20–21 March 2000, M'bé, Côte d'Ivoire. WARDA, Côte d'Ivoire. 95.

WARDA (2001b) *Summary of WARDA/NARS Task Forces Activities 1991–1997*. WARDA, Côte d'Ivoire, 218.

WARDA (2001c) *WARDA Strategic Plan 2001–2010*. Working Draft. WARDA Côte d'Ivoire, 36.

WARDA (2004a) *The Medium Term Plan 2005–2007: Charting the Future of Rice in Africa*. WARDA, Côte d'Ivoire, 76.

WARDA (2004b) *Strategic Plan 2003–2012*. WARDA. Côte d'Ivoire, 56.

WECARD/CORAF (1999) *Strategic Plan*. Draft, June. CORAF, Dakar, Sénégal.

9 Highlights of the Evolution of Priority Assessment and Targeting at the International Center for Maize and Wheat Improvement (CIMMYT)

JOHN DIXON AND ROBERTO LA ROVERE*

Abstract
The International Maize and Wheat Improvement Center's (CIMMYT) experience with priority assessment and targeting has co-evolved with the advances across the international agricultural research system. A wide range of CIMMYT *ex post* assessments provide a valuable knowledge base of returns to past research which reduces the risks of poor research choices; but their value is increased by *ex ante* impact assessments. One milestone for the allocation of resources was the development of a Resource Allocation Tool ('RAT'), used at CIMMYT during its strategic planning in 2003/04. The RAT served as a starting point to address complex resource allocation questions in quantitative ways with hard data. These experiences with priority assessment and targeting approaches, methods, criteria and data were reviewed in a workshop at the CIMMYT headquarters in Mexico in August 2005. For the purposes of identifying optimal allocations of resources to maize and wheat research, by region, by discipline (e.g. genetic resources, breeding or agronomy) and by challenge/constraint (such as drought, biotic stress or value-added), a systematic step-wise process was outlined. This included the choice of criteria and methods depending on the specific demands, scope, level of aggregation and availability of resources.

Keywords: Priority assessment, targeting, resource allocation, maize improvement, wheat improvement, agronomy

Introduction

The International Maize and Wheat Improvement Center's (CIMMYT) experience with priority assessment and targeting, in the sense of identifying key

* Impacts targeting and assessment unit, CIMMYT. The authors appreciate input from past and current colleagues. The contents of this chapter reflect the opinions of the authors and do not imply any policy on the part of CIMMYT or other CGIAR Center.

farming zones and populations on which the research is focused, has co-evolved with the wider experience of the Consultative Group on International Agricultural Research (CGIAR) system. Even with a strong focus on two major cereals, a central challenge has been the wide variety of agricultural research issues associated with the CGIAR and CIMMYT missions. This chapter traces the evolution of priority assessment and targeting in CIMMYT within the context of the shifting landscape of food systems, international and national research capacity and research funding, and highlights a number of milestones. This first section introduces some of the key changes in the context for priority assessment at a CGIAR Center. The second section provides an overview of the distribution of maize and wheat systems and National Agricultural Research Systems (NARS) capacity in developing countries, and highlights early experience (until 2002) with priority assessment and targeting in CIMMYT. In the third section, the Resource Allocation Tool (RAT), which provided information for the formulation of the Seeds of Innovation strategy in 2003, is described. The reader is brought up to date in the fourth section, which considers some highlights of priority assessment during the following five years, and the fifth section leads to a discussion of some recent initiatives in CIMMYT to reinforce the underlying databases for priority assessment. Conclusions are presented in the final section.

As noted by the World Bank (2003), the vision and priorities of the CGIAR system have evolved over time. In the mid- to late 1990s the CGIAR shifted its overarching goal from global food security (the 'pile of rice') to poverty reduction, through food security, resource conservation and management, policy enhancement and strengthening national agricultural research systems (Herdt, 2001). Agricultural research is a long-term activity and resource allocation decisions have long-term consequences for research productivity, outcomes and social impacts. If priority assessment and targeting for efficient resource allocation are important when funds are plentiful, they are critical in an environment of tightening resources in the CGIAR Center (Byerlee, 2000). Research resource allocation is conditioned by the relationship among Center, NARS and donors. Recent surveys of donors (e.g. Raitzer and Kelley, 2008) show a strong demand for more information related to impacts and priorities in agricultural research, especially knowledge which would underpin strategic decisions on types of agricultural research investment (e.g. food cf. feed maize, productivity cf. maintenance research, germplasm development cf. crop system management cf. crop value chain development). In a wider context, the emphasis in the evaluation and impact assessment profession is swinging from mainly *ex post* impact assessment (for accountability and estimation of historical returns to research) towards more *ex ante* impact assessment (for comparing future benefit streams from different research activities). Generally these *ex ante* impact assessments concentrate on a sub-sector or a research product, e.g. maintenance research on wheat rust.

Methods of assessment have also evolved. Alston *et al*. (1995) review a number of relevant priority assessment methods, ranging from economic welfare estimated by producer and consumer surplus to scoring and congruence approaches. Congruence and scoring models were popular during the 1980s and 1990s due to their simplicity and low cost, but subsequent efforts have been, at times, more sophisticated. A key determinant of the choice of approach for priority assessment is the time frame and available resources, notably the availability of experienced researchers to estimate parameters related to research outputs, impact pathways and impacts, and to reflect on, and act upon, the results. In many cases, such assessments are most valuable when supplemented by systematic discussion of the 'big picture' concerning production and consumption trends.

Systematic priority assessment requires consideration of many aspects by research managers and scientists. The expected benefits of research depend on many factors, including the probability of research success within a specified time period (often 5–25 years for crop improvement research), probable crop production impact (including crop area, yield increase and reduction of inputs) and probable 'people-level' impact (including improved food consumption and numbers of people who are lifted out of poverty) – with suitable accounting of spillovers to non-target areas and other indirect benefits on natural resources and non-agricultural sectors. Careful consideration of impact pathways is essential, including: the years to seed release, adoption, probability of widespread adoption conditioned by effectiveness of partners and appropriateness of institutions and policies, and probability of yield increments and other impacts following adoption. Within Centers, such as CIMMYT, crop research and knowledge sharing can be conceptualized as value chains from the genebank to farmers' fields (and ultimately to the consumer) through some or all of the steps of conservation of genetic resources, strategic germplasm enhancement (or pre-breeding), adaptive breeding, seed multiplication and dissemination, crop systems management, value chain institutions/policies improvement, supported throughout by impact assessment, targeting, knowledge sharing and partnership enhancement.

Priority assessment and targeting are closely associated with planning and *ex post* and *ex ante* impact assessment (Alston *et al.*, 1995; Janssen and Kissi, 1997; Bigman and Fofack, 2000). Two recent trends reinforce the complementarity among priority assessment, impact assessment and targeting. First, there is a tendency towards supplementing the accountability function of *ex post* impact assessment with an internal learning function (see Chapter 2, this volume, for example). Second, as noted above, the emphasis is shifting from *ex post* impact assessment towards scenario analysis and *ex ante* impact assessment (such as in Chapter 11, this volume). In *ex ante* assessment, it is recognized that the past performance and impacts are an imperfect mirror of potential future outcomes and impacts, and therefore, expert judgement is required. Preferably such expert judgements are made by biological or social scientists who have personal familiarity with the research area and target systems. Increasingly, priority assessment exercises combine analytical methods with expert judgement and stakeholder consultations, such as the identification of the new CGIAR system priorities in 2005 (described in Chapter 12, this volume). Such composite approaches are useful in the context of strengthening partnerships between the CGIAR system and NARS, and are also potentially valuable for attracting donor investment to key research areas.

Context and Early Development of Priority Assessment and Targeting in CIMMYT

Maize and wheat are major sources of food calories in more than 150 countries. In the spirit of congruency analysis, Table 9.1 lists the current areas of maize and wheat in the 50 largest producing countries in developing regions, classified into four groups according to the national areas of production. This represents a frame of reference for ascertaining the relative importance of the potential major target areas in the developing world for maize and wheat system improvement. The Table also provides an indication of current research capacity, measured in numbers of scientists of the NARS. Of course, the growing capacity of many NARS since 1960 has re-engineered the partnerships between NARS and the CGIAR system, in general, and for CIMMYT, in particular.

In general terms, priority assessment in CIMMYT is characterized by a number of steps which are implemented with varying degrees of detail (see also Alston *et al.*, 1995):

1. Identify the priority assessment leadership team, ideally cross-disciplinary or cross-stakeholder.
2. Identify beneficiary needs and associated problems/opportunities.
3. Define research objectives, research alternatives and priority-setting criteria.
4. Estimate benefits of research alternatives.
5. Conduct sensitivity analysis against priority-setting criteria.
6. Ultimately, synthesize results, review the process and identify the final choice of the priority research activities.

Two important decisions relate to the level at which the prioritization takes place and to the weighting of different criteria. Common priority assessment criteria from other exercises, such as those described in other chapters of this book, include: feasibility and available expertise; fundability; efficiency; equity (distinguishing the poor and the very poor); relevance to client needs; sustainability and existence of feasible delivery pathways; and internationality. For all criteria, expert judgements from scientists are an important source of information. The trade-offs between criteria (or objectives) should be identified, so that the relative weights of the criteria can be established early in the process and included in the framework. Prioritization of projects can include aggregation of the criteria into a single score using an appropriate weighting scheme.

CIMMYT has four decades of tacit and explicit experience with priority assessment and targeting. In the early 1960s, Dr Norman Borlaug undertook, for the UN Food and Agriculture Organization (FAO), one of the first major assessments of the potential for developing improved wheat in North Africa and Asia (Hessen, 2006), which laid the basis for the establishment and identification of the strategies of CIMMYT. During the 1970s, expert judgement of leading scientists, management and the Board was the key determinant of research priorities (and continues to this day to play a major role). As the maize and wheat research programmes grew in size and complexity during the 1980s,

Table 9.1. Characterization of 50 largest maize- and wheat-producing developing countries.

Area maize and wheat (million ha) <number of countries>	Countries (million ha)	Maize area (million ha) <yield> (t/ha)	Wheat area (million ha) <yield> (t/ha)	GDP ($000 PPP/cap)	HPI (%)	Agricultural Population (million)	Research capacity (approx FTE>s)
>10	Chi – 47, Ind – 34, Bra – 15, Kaz –12, Tur – 10	9.2	14	5.36	16	289	5107
<5>		<3.3>	<2.3>				
2–9.9	Pak – 9, Arg – 8, Mex – 8, Ukr – 8, Ira – 7, SAf – 4,	2.3	2.7	4.65	28	30	1316
<13>	Nig – 4, Ins – 3, Mor – 3, Eth – 3, Phi – 3, Tan – 2, Alg – 2	<2.6>	<2.1>				
1–1.9	Egy, Syr, Afg, Irq, Mal, Uzb, Ken, Nep, Con,	0.8	0.7	2.56	32	15	416
<14>	Moz, Zim, Tha, Ang, Tun	<2.2>	<2.0>				
0.5–0.99	Vie, Turk, Par, Uga, Bgd, Gha, Ben, Col,	0.4 <2.3>	0.2 <2.3>	3.90	19	12	682
	Zam, Aze, Ven, Gua,						
<18>	Per, Kor(Dem), Chi, Sau, Cam, Ecu						
Average <50>		2.0	2.4	3.89	25	45	1009

Data drawn from RAT, FAOSTAT (2004) and UNDP (2006); standard UN country abbreviations: FTE – Full Time Equivalents (researchers).

agricultural economists began to work with agronomists and breeders to assess impacts of improved varieties and to identify crop improvement priorities. For example, Heisey *et al.* (1990) identified research priorities for high-altitude maize farming systems in Swat, Pakistan, based on data from diagnostic surveys, whereas Shumba *et al.* (1990) used a yield gap framework to identify maize and groundnut research priorities in Zimbabwe. In relation to simpler farming systems, Bell *et al.* (1994) identified crop management research priorities in the Yaqui Valley and Meng *et al.* (2000) analysed the economics of wheat production in Kazakhstan and identified local priorities for wheat research. The conduct of diagnostic studies to support research priority assessment absorbed a significant proportion of agricultural economics research capacity during the 1980s. At the global level, analyses of maize and wheat potentials can also be found in the well-known series of publications on maize facts and trends (e.g. Lopez-Pereira and Morris, 1994; Pingali, 2001) and wheat facts and trends (e.g. Byerlee and Moya, 1993; Ekboir, 2002; Dixon *et al.*, 2008).

In the post-Green Revolution era, one of the major broad priority assessment questions has been the relative pay-offs to investing in breeding for marginal compared with better-endowed environments (Byerlee and Traxler, 1995; Lantican *et al.*, 2003). An early analysis at the national level is provided by Byerlee and Hussain (1993), drawing on farming systems surveys in different zones of Pakistan. Such comparisons need to take account of potential spillovers as varieties developed for one environment can spread to other environments (Alston, 2002). In the case of maize and wheat improvement, the additional benefits generated from spillovers were substantial, especially with improved wheat lines developed through shuttle breeding and characterized by wide adaptability (Braun *et al.*, 1996).

Targeting of breeding to agroecologies and farming populations has been a major challenge for maize and wheat breeders. During the 1980s CIMMYT wheat breeders formalized the notion of breeding for geographic environments with similar agroecologies. Wheat 'mega-environments' have been defined which guide the targeting of breeding and seed system development (Braun *et al.* (1996) relate breeding for wide adaptation to the mega-environments for wheat production). In southern Africa, cluster analysis of maize trial data identified eight distinct maize environments, distinguished by growing season precipitation and temperature, which have been delineated and mapped (Hodson *et al.*, 2002; Setimela *et al.*, 2002).

The RAT

During 2003, CIMMYT conducted a formal Center-wide priority assessment exercise, in the context of the search for a new vision and modern research thrusts. The approach combined intensive internal and external consultations with a congruency analysis using the RAT. The approach and results were debated among scientists and contributed to the Center's vision, strategy and priorities set forth in 'Seeds of Innovation' (CIMMYT, 2004), which embraced, *inter alia*, maize and wheat genetic resources and diversity, nutritional value markets, resource-conserving technologies and knowledge sharing.

The RAT was developed in May 2003 by the Priority Setting Working Group, comprising scientists from maize improvement, wheat improvement, crop systems management and economics, as input into the strategic planning process. The RAT is a congruency model grounded on the assumption that resources should be allocated across research activities in proportion to the estimated values of the commodities associated with the research outputs.[1] The database for the RAT comprises: production and consumption data; value chain constraints and opportunities on maize and wheat; and comparative advantage for CIMMYT cf. other research providers, and poverty data, predominantly at the national level for the year 2000. In addition, a survey of staff and NARS scientists provided information on research needs, research capacity, alternative suppliers and other information related to impact pathways.

The RAT takes as its starting point the distribution of maize and wheat consumption in developing countries. Although the baseline uses data of the year 2000, projected consumption data up to and including 2020 can be substituted to explore the implications of future scenarios for Center research priorities. The model is based on consumption data, in recognition of the fact that a significant proportion of the maize and wheat consumed by the poor in developing countries is imported. RAT distinguishes three factors of generic relevance: the importance of maize and wheat in human livelihoods; the incidence of poverty; and the strength of partners and presence of alternative suppliers (to CIMMYT) of research knowledge and associated services, e.g. capacity building. Two national-level indices have been developed for each factor.

The two indices for the first factor, the importance of maize and wheat, are:

$$P * \sqrt{a * \left(\frac{b}{c}\right)}$$

where p = agricultural population, a = consumption crop, b = crop area, c = total cropped area; and

$$A * Cp,$$

where A = maize or wheat area, Cp = per capita consumption of maize or wheat.

The second factor, national poverty, is represented by alternate indices:

$$(HPI)^2,$$

where HPI = human poverty index, derived from United Nations Development Programme (UNDP); and

$$1/GDPpc,$$

where $GDPpc$ = gross domestic product per capita in the agricultural sector.

The third factor, partner strength, is represented in contrasting ways:

$$(Snp + Sp) * Ic,$$

[1] The model description is derived largely from internal working papers and presentations prepared by M. Morris and D. Hodson, respectively, the leader and a member of the Priority Setting Working Group.

where *Snp* = strength of public, NARS organizations, *Sp* = strength of private companies, *Ic* = CIMMYT's likely impact derived from the survey; and

HPI,

where *HPI* = human poverty index.

The choice of the human poverty index (HPI), reflects the association of research capacity with general economic development and the potential value of composite indices developed, reviewed and updated by major development organizations such as FAO and the World Bank.

For the purpose of supporting the formulation of the 'Seeds of Innovation' strategy in 2003/04, different runs of the RAT model were implemented for assessing the optimal allocation of Center research resources, corresponding to the eight different combinations of indices on the importance of maize or wheat, level of poverty and the strength of partners. Where expert judgements were required to determine parameter values to produce input for the strategic discussions, the working group used either consensus or averaged estimates of members. A set of illustrative results of the RAT, averaged across different runs, are shown in Tables 9.2. and 9.3. The first shows an optimal current allocation across regions and crops under the assumptions related to congruency: the global allocation can be compared with the existing allocation, in the early 2000s, of 59% of research resources to maize and 41% to wheat. It is noteworthy that the IMPACT model projections used for the 2020 production scenario indicate an expansion of the importance of maize, and thus, RAT results suggest an

Table 9.2. Illustrative results of RAT: target current research resource allocations (%).

Region	Maize	Wheat	Total
Sub-Saharan Africa	30	2	32
Latin America	4	0	4
East, South and South-east Asia	11	42	53
Central and West Asia, North Africa	1	12	13
Developing countries	45	55	100

RAT estimates.

Table 9.3. Illustrative results of RAT: target future research resource allocations (%).

	Maize	Wheat	Total
Sub-Saharan Africa	43	3	46
Latin America	3	0	4
East, South and South-east Asia	8	30	38
Central and West Asia, North Africa	1	11	12
Developing countries	55	45	100

RAT estimates

increase in the relative share of Center resources to be allocated to maize in the future. Of course the assumptions underlying RAT, of which many are noted above, as well as the choice of indices, are critical when interpreting these results. In addition, the RAT incorporates a set of variables that allow sensitivity analysis to be performed in four critical areas of interest: grain price ratio; food versus feed; CGIAR Center mandates; and potential for spillovers, especially in the case of large NARS. The approach to each is discussed in the following paragraphs.

Grain price ratio

The RAT analysis can be conducted in terms of physical units of maize and wheat consumption, or in terms of the value of maize and wheat consumption. Sound arguments can be made to justify either approach. If the analysis is conducted in terms of value, however, it is necessary to assign prices to maize and wheat. Ideally, local consumer prices would be used, but since these are not available for all countries, the RAT analysis is based on international reference prices for maize and wheat. Since international reference prices change through time, a variable was incorporated to allow changes in the ratio of maize to wheat prices. A sensitivity analysis was conducted for the maize/wheat price ratio, using values of 0.70, 1.00 and 1.42.

Food versus feed use

Given CIMMYT's mission to serve the poorest of the poor, it was argued that maize and wheat destined for human consumption (food use) should be assigned higher priority than maize and wheat destined to be fed to animals (feed). In this case, the feed discount factor which was used in 2003 was 0.42, reflecting the shorter impact pathways and wider impact poverty reduction 'footprint' of maize as a staple food – while not ignoring the important indirect poverty-reducing effects of animal fattening, dairy and other maize feed-based income generation activities.

Partners

Based on the survey results, factors were estimated for the relative strength of the public sector versus the private sector. Alternate CGIAR suppliers of improved maize and wheat systems, respectively, the International Institute for Tropical Agriculture (IITA) and the International Center for Agricultural Research for the Dry Areas (ICARDA), were represented by discounts on consumption of maize or wheat in the respective regions.

Potential for spillovers

Major NARS, including those of Brazil, China and India, are important partners for CIMMYT and account for a significant portion of the area of maize and wheat in developing countries. In practice, the potential for capturing research spillovers is much larger within or beyond these three countries than in other maize and wheat producing countries. For this reason RAT allows adjustments for such large NARS.

As noted above, the results derived from the RAT are not strict resource allocation parameters. The actual allocation of resources by region will depend on a number of other considerations. In relation to research efficiency, the allocation of research resources for a specific region does not mean that the research should take place within that region. A pertinent example is the Mexico-based breeding programme for wheat improvement: although breeding takes place in Mexico, advanced wheat lines are deployed in many other regions, as far afield as East and South Asia, CWANA, sub-Saharan Africa and South America. Another consideration relates to correspondence between the levels of analysis and the availability of funding. Because CIMMYT is not funded entirely through unrestricted core funds, allocation decisions are constrained by the donor earmarking of research funds for specific regions, purposes outputs and target populations. Nevertheless, RAT results may facilitate the mobilization of resources for particular areas of high-priority research.

Reflecting 'Seeds of Innovation' in Research Plans and Processes and Implications for the Future

In addition to the RAT results discussed in the previous section, NARS priority needs were identified through a survey. These are reported for maize and wheat in Tables 9.4. and 9.5. As would be expected, large NARS had greater capacity for pre-breeding, whereas smaller NARS placed more emphasis on germplasm and especially varietal development. Both component agronomy and systems agronomy were rated as important, and the higher scores for systems agronomy are noteworthy. Other high priorities included impact assessment and maize and wheat policy analysis. There is a significant degree of congruency with other assessments of research priorities, for example Kosina *et al.* (2007) for wheat or Gibbon *et al.* (2007) for maize. Not only does such information help to guide decisions on internal resource allocations, but also has the potential (not yet exploited fully) to redirect investment in international agricultural research to higher pay-off themes through better-informed dialogues and debates with donors.

The period from 2004 to 2008 was marked by several shifts in the frame of reference and organizational structures for priority assessment. From the adoption of the 'Seeds of Innovation' strategy in 2003, the various programmes identified priorities during annual planning meetings for the preparation of the medium-term plan (MTP) during 2004. An example of formal priority assessment occurred during the 2004 annual meeting of the CIMMYT Intensive Agro-Ecosystems Programme. Scientists developed a set of weights and priorities for purposes, results and activities in order to develop the programme MTP logframe. In this case, there were no large adjustments in priorities between

Table 9.4. Maize research needs by NARS group.

Area of Maize and Wheat (million ha) <number of Countries>	Countries (million ha)	Pre-breeding	Germplasm development	Variety development	Seed systems	Component agronomy	Systems agronomy	Postharvest	Impacts priority setting	Policy
>10 <5>	Chi – 47, India – 34, Bra – 15, Kaz – 12, Tur – 10	3.4	4.2	4.0	3.0	3.0	4.0	3.5	3.6	3.3
2–9.9 <13>	Pak – 9, Arg – 8, Mex – 8, Ukr – 8, Ira – 7, SAf – 4, Nig – 4, Indo – 3, Mor – 3, Eth – 3, Phi – 3, Tan – 2, Alg – 2	2.8	3.6	3.7	3.7	3.2	3.6	3.4	3.9	3.8
1–1.9 <14>	Egy, Syr, Afg, Irq, Mal, Uzb, Ken, Nep, Con, Moz, Zim, Tha, Ang, Tun	2.4	3.5	4.2	3.9	3.4	3.8	3.3	3.8	3.4

Data drawn from RAT survey; scoring 1 (low need) to 5 (great need).

Table 9.5. Wheat research needs by country group.

Area of Maize and Wheat (million ha) (no. countries)	Countries (million ha)	Pre-breeding	Germplasm development	Variety development	Seed systems	Component agronomy	Systems agronomy	Postharvest	Impacts priority setting	Policy
>10 <5>	Chi – 47, India – 34, Bra – 15, Kaz – 12, Tur – 10	4.0	4.2	3.6	3.2	2.8	4.0	3.2	3.6	3.8
2–9.9 <13>	Pak – 9, Arg – 8, Mex – 8, Ukr – 8, Ira – 7, SAf – 4, Nig – 4, Indo – 3, Mor – 3, Eth – 3, Phi – 3, Tan – 2, Alg – 2	3.1	3.5	3.5	2.9	3.1	3.3	2.0	3.5	3.3
1–1.9 <14>	Egy, Syr, Afg, Irq, Mal, Uzb, Ken, Nep, Con, Moz, Zim, Tha, Ang, Tun	2.5	3.2	3.5	2.8	3.0	3.0	2.1	3.1	2.8

Data drawn from RAT; scoring 1 (low need) to 5 (great need).

current activities and desirable future activities, because the programme already had made very substantial investments in areas identified in 'Seeds of Innovation'. Other programmes used less formal methods of prioritization and planning to incorporate adjustments indicated by Seeds of Innovation. In relation to the unrestricted budget, annual allocations to each of the eight MTP projects are determined by averaging the votes of management committee members for each MTP project with special additional allocations to some specific projects.

A more detailed priority assessment framework that took explicit account of impact pathway length (in years), probability of research success and projected impacts on food security and poverty with scoring linked to quantified research processes and impacts, was desigined although not ultimately implemented. The proposed approach involved broadly quantified scores related to key dimensions of success probability, impact pathway duration and difficulty, and expected food security and poverty reduction impacts. Such an approach has the advantage of forcing explicit judgements on key bottlenecks in the research-to-impact chain and fostering examination of the assumptions about delivery pathways. Although scoring models are open to manipulation of the weighting schemes, other projection, *ex ante* impact assessment and congruency models are not immune to this potential weakness – it has been observed how quickly scientists learn to adjust key model assumptions and parameters to generate 'favourable' results. Utilizing the impact pathway as the backbone of the priority assessment framework focuses managers' attention on the lag time to impact – although in practice some managers do not accept the implications of, ceteris paribus, relatively lower allocations of resources to activities with relatively longer impact pathways. From an overall perspective, such priority assessment frameworks confer a variety of advantages including simplicity, transparency, i.e. ease of comprehension, and, above all, low cost. It should be noted that the majority of the Center budgets are derived from restricted projects whose research objectives and outputs are determined through consultation with partners and investors and/or competition with other research organizations within the frame of the Center MTPs, and are sometimes excluded from such specific priority assessment (although as mentioned above the analyses could in principle support dialogues with donors about priorities).

During August 2005, an internal workshop on priority assessment and targeting was organized in CIMMYT with the participation of scientists from all the programmes. Participants concluded that priorities should be explicitly established at all management levels in CIMMYT, for the Center, the programme, the MTP project, the MTP output and the research project, in line with the principle of subsidiarity. The frame of reference for priority assessment at the Center level would be provided by Seeds of Innovation, the six research programmes and the 11 MTP projects; at restricted project level, the frame of reference would be the specific project's objectives and outputs. From the point of view of supporting an impact assessment orientation and culture, this workshop complemented two other impact assessment workshops during the year, namely, a CIMMYT impact assessment workshop in May 2005, which launched an impact assessment learning platform, and a Participatory Research and Gender Analysis (PRGA) CGIAR system-wide program workshop on impact assessment hosted by CIMMYT in October 2005 (La Rovere *et al.*, 2008). During 2006, the frame of reference for priority assessment changed substantially, with the reduction in the number of MTP projects from 11 to the current eight, the reduction from six research programmes to four research programmes and units, and the adoption of a business plan which set forth nine broad 'flagships'. Internal discussions also identified the need for better understanding of the implications of impact pathways for research prioritization, and thus an impact pathways workshop was organized with participants representing the eight MTP projects. This workshop clarified concepts related to outputs, outcomes, impacts; identified the often-complex networks of actors, which converted the Center outputs to impacts on the ground; and mapped networks and impact pathways for each of the eight projects.

In general, a poverty orientation is being more strongly reflected in research targeting. One assessment tested methods for targeting crop improvement to poor farmers (Bellon *et al.*, 2003). Interactive poverty maps were constructed using poverty and agroclimatic data for Mexico. Using multivariate classification and cluster analysis, biophysical data related to maize productivity and to rural poverty were synthesized. It was found that most formal maize improvement trials are outside the core areas of rural poverty and little evidence could be found of direct spillovers into the areas of acute poverty. Through this combination of agroclimatic and rural poverty data, improved targeting of crop-breeding efforts to meet the demands of resource-poor farmers can be attempted.

The geographic dimensions of poverty and food security, particularly through the use of spatial poverty databases for targeting, are a growing theme in priority assessment activities. A good example of their use for sub-regional priority assessment can be found in the Rice–Wheat Consortium for the Indo-Gangetic Plains (RWC). Erenstein *et al.* (2007) applied a spatial mapping methodology for the Indo-Gangetic Plains (IGP) for guiding priority assessment and targeting in the RWC. The approach incorporates livelihood and poverty considerations. It draws from secondary data and uses 18 quantitative, spatially explicit variables, which serve as indicators of poverty levels based on livelihood capitals of the sustainable livelihoods approach: natural, social, human, physical and financial. The livelihood assets approach to measuring poverty complements the more conventional monetary approach of assessing poverty. The study details the methodology used to generate spatial poverty maps at the district level for composite indicators for each livelihood asset and an overall livelihood asset index. Each individual asset index shows statistically significant variation between transects of the IGP, which cover the states of Punjab, Haryana, Uttar Pradesh, Bihar and West Bengal. The overall livelihood asset index showed a significant and strong negative correlation ($r = -0.55$) with the more conventional monetary measure for assessing poverty. This database needs to be complemented by a better understanding of previous development and research experiences to shed more light on farmers' livelihoods and strategies in the IGP and the impact pathways and networks that link research outputs (technology and knowledge) to farm household adoption of improved varieties and practices, and to ultimate impact.

Another major assessment demonstrated the feasibility of participatory approaches to data collection and research priority assessment on a regional scale (Gerpacio and Pingali, 2007). Rapid Rural Appraisal (RRA) and Participatory Rural Appraisal (PRA) techniques were used in conducting village-level and farmer-group surveys for the detailed characterization of upland maize production environments and systems in China (Meng *et al.*, 2006), India (Joshi *et al.*, 2005), Indonesia (Swastika *et al.*, 2004), Nepal (Paudyal *et al.*, 2001), the Philippines (Gerpacio *et al.*, 2004), Thailand (Ekasingh *et al.*, 2004) and Vietnam (Thanh Ha *et al.*, 2004). The survey findings were fed into country-level national maize technology research and development (R&D) priority-setting workshops, which developed R&D priorities for specified maize production environments and market orientation. Across the region, addressing the problem of drought in the rain-fed lowland-commercial, rain-fed upland-semi-commercial and rain-fed commercial production environments would provide the highest technical returns to maize R&D investments in Asia. Downy mildew, stem borers, leaf blight, stalk rot, soil erosion/landslides, soil micronutrient deficiency, waterlogging, lack of capital or poor access to low-interest credit, poor agricultural extension/technology transfer services and poor access to input and output markets commonly figured in the priority constraints lists as well. Socio-economic and policy-related constraints were estimated to affect up to 180 million rural poor people across Asia and their alleviation could improve maize productivity by at least 18%.

It is interesting to consider the degree to which CIMMYT funding has evolved along the lines indicated by the RAT 2003 results. One of the conclusions in 2003 was that CIMMYT should invest more resources on maize as opposed to wheat, and this has since come about. Second, the RAT suggested that CIMMYT was over-invested in Latin America in 2003; and since then the level of resources has reduced (compared with East, South, and South-east Asia).

What does this experience suggest for the future? The methods, criteria, approaches and data for priority assessment activities at CIMMYT have evolved, largely based on in-house experience within the frame set by 'Seeds of Innovation' (CIMMYT, 2004) and the business plan (CIMMYT, 2006). Some of the key unresolved issues that have emerged from the experience are: (i) the optimal scale for priority assessment; (ii) identification of multiple criteria and weights; (iii) the critical lack of resources earmarked for priority assessment; (iv) the need for multidisciplinary team involvement throughout the process; (v) the poor quality and coverage of available data; and (vi) the engagement of donors in the priority assessment process.

There are a variety of new challenges for priority assessment and targeting. Perhaps one of the most profound is the nexus between climate change and risk. In terms of shifting long-term averages, Hodson and White (2008) report the halving of the favourable (mega-environment 1) wheat area in the rice–wheat farming system in South Asia by 2050 because of heat stress; and the prospects for expansion of wheat in northern latitudes including Siberia because of increasing precipitation and temperature. More recently, the risk of mid-season drought in rain-fed maize is being simulated and mapped as an input to drought-tolerant maize breeding and seed deployment strategies in sub-Saharan Africa. Perceptions concerning risk, including climate risk, are of course a key factor in farmers' decisions. These considerations are of particular relevance to the adoption, replacement, management and impact of improved maize and wheat varieties, especially since current models suggest that climatic variability may well increase in many locations affected by climate change.

Priority assessment methods are relatively well developed for individual productivity-enhancing and cost-reducing production technologies. Although assessments are fewer in number, the concepts for impact assessment and priority assessment for maintenance research, where yield gains are defended against biotic or abiotic stress have been demonstrated. Similarly, the benefits of quality-enhancing research can be estimated for quality traits which are reflected in grain or stover/straw market prices. However, a majority of assessments have avoided the difficulties of positive or negative interactions of technologies, such as improved varieties and associated improvements in crop management. Moreover, the valuation of less-observable traits such as high lysine and tryptophan in quality protein maize poses a number of challenges (see Lauderdale, 2000). Moreover, with urbanization, improved market access and greater attention to food safety, priority assessment and targeting methods require further elaboration for assessing crop improvements with impacts focused primarily along the value chain instead of at producer level, e.g. varieties with reduced susceptibility to mycotoxins. Priority assessment related to food safety would benefit from linkages with the food-chain risk assessment literature.

Three further challenges merit mention. First, as noted in the preceding discussion, donors and decision makers are increasingly demanding a wider set of benefit metrics for the assessment of research, of which the 'livelihood' metrics is one part. Second, as Dixon *et al.* (2007) suggest, the full benefits of crop improvement have generally been underestimated because secondary benefits along input and output value chains and in the local non-farm rural economy have often been ignored. Although information on the magnitudes of multipliers stimulated by cereal intensification is scarce, both first principles and a few studies suggest that the secondary effects from cereal intensification may often be substantial. Third, the priority assessment community has not come to grips with the implications of price variability of the magnitude of recent years. In the recent past, commodity price ratios, e.g. the maize/rice price ratio, as well as input/output price ratios, e.g. fertilizer/grain price ratios, have changed sufficiently to call into question earlier findings on the relative priorities for crop improvement research.

Data Sources for Priority Assessment and Targeting

In reality, the choice of approach to priority assessment is often constrained or even determined by the availability of existing data. In many cases, much information is available from a variety of sources: published data, research reports, expert knowledge of research and extension workers, and key informants including farmers – although skill is required to synthesize such disparate and diverse data. The following areas of data, grouped by scale, can be particularly important for priority assessment: global, notably potential production/consumption and their trends, constraints and opportunities; national, as incorporated into the RAT; sub-national, especially farming systems, production and value-chain constraints; and local village/household data, including livelihoods, off-farm income and vulnerability. Types of data for which adequate detail is often missing include: migration demographics, farming systems, diversification options, research constraints, target group profiles, and costs of different research options. Both quantitative and qualitative data (e.g. expert opinions) can be useful. A basic inventory of current available data should be put in place for the priority assessment teams, with regular updates and identification of gaps at different levels. In this sense, the preperation for and conduct of priority assessment is an ongoing and iterative process.

Conclusions

There is a continuing stream of priority assessment and targeting activities in CIMMYT, which can be cast in a multi-step framework. Over the decades, informal methods have been more widely applied than formal quantitative priority assessment methods. Given budget realities, low 'cost' methods have dominated the CIMMYT experience; however, there is recognition of the value of formal quantitative methods such as the RAT and the need for a broadening of the criteria set to include social and environmental measures in the appraisal of priorities. RAT was a milestone in the development of priority assessment and targeting in CIMMYT, and the value of updating and enriching RAT has been recognized. Although the use of methods that focus more on the external environment can be appropriate for dealing with complex and risky systems and cost-effective given the current funding landscape, this comes at some cost to the transparency of the results. *Ex post* and *ex ante* assessments provide an important knowledge base for scientists and research managers which enables managers to use, in principle, disaggregated congruency and partially quantified impact scoring models. The optimism of scientists concerning their own research areas and the short horizon for research planning within CGIAR Center presents a challenge to the institutionalization of priority assessment. There is also a tendency for some managers to expect impact assessment and priority assessment to generate the 'right results' for use in the promotion of specific interests rather than as a tool to adjust the overall research portfolio in consultation with donors, for maximum impact. Given that donor and Center research goals are often similar, the results of priority assessment could in principle be very useful input to research resource mobilization dialogues.

Notwithstanding the clear evidence of varied, albeit, high rates of return to research for maize and wheat system improvement, the salience of priority assessment has declined as the proportion of restricted, objective-bound research funding has increased. The decline in attention to priority assessment has also been reflected in a world-wide reduction of the quality and quantity of available data for this purpose – a constraint which limits much economic and social analysis of agricultural development as well.

However, there is at CIMMYT renewed interest and demand for a structured and robust priority assessment process to focus and refine the CIMMYT strategy and business plan. There are also strengthened internal capacity and new external partnerships, which could take forward such assessments. Therefore, opportunities are emerging to re-establish databases which serve priority assessment and targeting and to make priority assessment a core activity to inform management of pay-offs across different research portfolios.

References

Alston, J.M. (2002) Spillovers. *Australian Journal of Agricultural and Resource Economics* 48 (3), 315–346.

Alston, J.M., Norton, G.W. and Pardey, P.G. (1995) *Science Under Scarcity: Principles and Practice for Agricultural Research Evaluation and Priority Setting*. Cornell University Press, Ithaca, New York.

Bell, M.A., Raab, R.T. and Violic Martinovik, A.D. (1994) *Setting Research Priorities for Agronomic Research: A Case Study for Wheat in Chalco, Mexico. Wheat Special Report (WPSR) 33*. CIMMYT, Mexico DF, Mexico.

Bellon, M.R., Hodson, D., Bergvinson, D., Beck, D., Martinez-Romero, E. and Montoya, J. (2003) Targeting agricultural research to benefit poor farmers: relating poverty mapping to maize environments in Mexico. *Food Policy* 30(5/6), 476–492.

Bigman, D. and Fofack, H. (eds) (2000) *Geographical Targeting for Poverty Alleviation: Methodology and Applications*. World Bank, Washington, DC.

Braun, H.J., Rajaram, S. and van Ginkel, M. (1996) CIMMYT's approach to breeding for wide adaptation. *Euphytica* 92, 175–183.

Byerlee, D. (2000) Targeting poverty in priority setting for agricultural research. *Food Policy* 25(4), 429–45.

Byerlee, D. and Hussain, T. (1993) Agricultural research strategies for favoured and marginal areas: the experience of farming systems research in Pakistan. *Experimental Agriculture* 29, 155–171.

Byerlee, D. and Moya, P. (1993) *Impacts of International Wheat Breeding Research in the Developing World, 1966–90*. CIMMYT, Mexico DF, Mexico.

Byerlee, D. and Traxler, G. (1995) National and international wheat improvement research in the post green revolution period: evolution and impacts. *American Journal of Agricultural Economics* 77, 268–278.

CIMMYT. (2004) *Seeds of Innovation: CIMMYT's Strategy for Helping to Reduce Poverty and Hunger by 2020*. CIMMYT, Mexico DF, Mexico.

CIMMYT. (2006) *CIMMYT Business Plan 2006–10: Translating the Vision of Seeds and Innovation into a Vibrant Work Plan*. CIMMYT, Mexico DF, Mexico.

Dixon, J., Hellin, J., Erenstein, O. and Kosina, P. (2007) U-impact pathway for diagnosis and impact assessment of crop iimprovement. *Journal of Agricultural Science* 145, 1–12.

Dixon, J., Braun, H.-J. and Kosina, P. (eds) (2008) *Wheat Facts and Futures – 2007*. CIMMYT, Mexico DF, Mexico.

Ekasingh, B., Gypmantasiri, P., Thong Ngam, K. and Krudloyma, P. (2004) *Maize in Thailand: Production Systems, Constraints, and Research Priorities*. CIMMYT, Mexico DF, Mexico.

Ekboir, J. (ed.) (2002) CIMMYT 2000–2001. *World Wheat Overview and Outlook: Developing No-till Packages for Small-scale Farmers*. CIMMYT, Mexico DF, Mexico.

Erenstein, O., Hellin, J. and Chandna, P. (2007) *Livelihoods, Poverty and Targeting in the Indo-Gangetic Plains: A Spatial Mapping Approach*. CIMMYT and RWC, New Delhi, India.

FAOSTAT. (2004) *Food and Agriculture Organization of the United Nations Statistical Database*. FAO, Rome, Italy. Available at: http://www.fao.org/corp/statistics/en/
Gerpacio, R.V. and Pingali, P.L. (2007) *Tropical and Subtropical Maize in Asia: Production Systems Constraints, and Research Priorities*. CIMMYT, Mexico DF, Mexico.
Gerpacio, R.V., Labios, J.D., Labios, R.V. and Diangkinay, E.I. (2004) *Maize in the Philippines: Production Systems, Constraints, and Research Priorities*. CIMMYT, Mexico DF, Mexico.
Gibbon, D., Dixon, J. and Flores, D. (2007) *Beyond Drought Tolerant Maize: Study of Additional Priorities in Maize*. GCP-CIMMYT, Mexico DF, Mexico.
Heisey, P.W., Ahmad, M., Stevens, E.J., Khan, K., Zeb, J. and Javed, H.I. (1990) *Diagnosing Research Priorities for Higher-altitude Maize-based Farming Systems in Swat*. PARC/CIMMYT Paper 90–3. PARC/CIMMYT, Islamabad (Pakistan).
Herdt, R.W. (2001) *Changing Priorities for International Agricultural Research*. CIMMYT Distinguished Economist Lecture 5. CIMMYT, Mexico DF, Mexico.
Hessen, L. (2006) *The Man Who Fed the World*. Durban House, Dallas, Texas.
Hodson, D. and White, J.W. (2008) Climate change – What future for wheat? In: Dixon, J., Braun H.-J. and Kosina. P. (eds) *Wheat Facts and Futures – 2007*. CIMMYT, Mexico DF, Mexico.
Hodson, D.P., Martinez-Romero, E., White, J.W., Corbett, J.D. and Bänziger, M. (2002) *Africa Maize Research Atlas (v. 3.0)*. CD-ROM publication. CIMMYT, Mexico DF, Mexico.
Janssen, W. and Kissi, A. (1997) *Planning and Priority Setting for Regional Research: A Practical Approach to Combine Natural Resource Management and Productivity Concerns*. Research Management Guidelines 4. ISNAR, The Hague.
Joshi, P.K., Singh, N.P., Singh, N.N., Gerpacio, R.V. and Pingali, P.L. (2005) *Maize in India: Production Systems, Constraints, and Research Priorities*. CIMMYT, Mexico DF, Mexico.
Kosina, P., Reynolds, M., Dixon, J. and Joshi, A. (2007) Stakeholder perception of wheat production constraints, capacity building needs, and research partnerships in developing countries. *Euphytica* 157, 475–483.
La Rovere, R., Dixon, J. and Hellin, J. (2008) *Institutionalizing Impact Assessment at CIMMYT*. ILAC Brief 18. ILAC, Rome, Italy.
Lantican, M.A., Pingali, P.L. and Rajaram, S. (2003) Is research on marginal lands catching up? The case of unfavourable wheat growing environments. *Agricultural Economics* 29, 353–361.
Lauderdale, J. (2000) *Issues Regarding Targeting and Adoption of Quality Protein Maize (QPM)*. Economics Working Paper 00–02. CIMMYT, Mexico DF, Mexico.
Lopez-Pereira, M.A. and Morris, M.L. (1994) *Impacts of International Maize Breeding Research in the Developing World, 1966–90*. CIMMYT, Mexico DF, Mexico.
Meng, E.C.H., Longmire, J., and Moldashev, A. (2000) Kazakhstan's wheat system: priorities, constraints, and future prospects. *Food Policy* 25 (6), 701–717.
Meng, E.C.H., Ruifa Hu, Xiaohua Shi and Shihuang Zhang. (2006) *Maize in China: Production Systems, Constraints, and Research Priorities*. CIMMYT, Mexico DF, Mexico.
Paudyal, K.R., Ransom, J.K., Rajbhandari, N.P., Adhikari, K., Gerpacio, R.V. and Pingali, P.L. (2001) *Maize in Nepal: Production Systems, Constraints, and Priorities for Research*. NARC and CIMMYT, Kathmandu, Nepal.
Pingali, P.L. (ed.) (2001) CIMMYT 1999/2000 *World Maize Facts and Trends. Meeting World Maize Needs: Technological Opportunities and Priorities for the Public Sector*. CIMMYT, Mexico DF, Mexico.
Raitzer, D.A. and Kelley, T.G. (2008) Assessing the contribution of impact assessment to donor decisions for international agricultural research. *Research Evaluation* 17(3), 187–199.
Setimela, P., Chitalu, Z., Jonazi, J., Mambo, A., Hodson, D. and Bänziger, M. (2002) Revision of maize mega-environments in the Southern African Development Community (SADC) region. In: Arnel R. Hallauer (ed.) *Book of Abstracts*. International Symposium on Plant Breeding. CIMMYT, Mexico DF, Mexico.

Shumba, E.M., Bernstein, R.H. and Waddington, S.R. (1990) Maize and groundnut yield gap analysis for research priority setting in the smallholder sector of Zimbabwe. *Zimbabwe Journal of Agricultural Research* 28 (2), 105–113.

Swastika, D.K.S., Kasim, F., Suhariyanto, K., Sudana, W., Hendayana, R., Gerpacio, R.V. and Pingali, P.L. (2004) *Maize in Indonesia: Production Systems, Constraints and Research Priorities*. CIMMYT, Mexico DF, Mexico.

Thanh Ha, D., Dinh Thao, T., Tri Khiem, N., Xuan Trieu, M., Gerpacio, R.V. and Pingali, P.L. (2004) *Maize in Vietnam: Production Systems, Constraints, and Research Priorities*. CIMMYT, Mexico DF, Mexico.

UNDP (2006) *Human Development Report 2006: Beyond Scarcity: Power, Poverty and the Global Water Crisis*. United Nations Development Programme, New York. Available at: http://hdr.undp.org/hdr2006/

World Bank (2003) *The CGIAR at 31: An Independent Meta-Evaluation of the Consultative Group on International Agricultural Research*. Washington, DC.

10 The International Center for Agricultural Research in the Dry Areas' (ICARDA) Experience in Agricultural Research Priority Assessment*

Kamil Shideed, Mahmoud Solh, Ahmed Mazid and Mazen El-Solh

Abstract

Past agricultural priority assessment exercises have often emphasized efficiency or productivity gain as a single objective. This emphasis is appropriate for commodity-based research, but can be problematic for non-commodity research, such as research that addresses natural resources or other cross-cutting topics. The International Center for Agricultural Research in the Dry Areas (ICARDA) research portfolio addresses multiple objectives, including equity (poverty alleviation) and environmental sustainability, which may involve trade-offs with short-term productivity gains. A key objective of ICARDA's research priority assessment is to lay the foundation for a regional partnership that facilitates consensus on agricultural research priorities to be addressed within a regional framework. The basic premise is that a regional process would strengthen partnerships among the National Agricultural Research Systems (NARS) and create opportunities for more efficient use of scientific and other resources to enhance the likelihood of impact. In 2002, ICARDA, in collaboration with regional organizations, implemented an extensive research priority assessment exercise for Central and West Asia and North Africa (CWANA). This exercise involved brainstorming meetings at subregional levels, a formal questionnaire that was widely distributed across the region to various stakeholders and a final regional consultation that brought together (national, regional and international) stakeholders with different backgrounds and mandates. Comparisons across research themes were conducted using a scoring approach with well-defined evaluation criteria and associated weights. This bottom-up approach resulted in identifying regional and subregional research priorities. The main purpose of the regional approach was to facilitate a holistic approach to fighting poverty through interregional and international cooperation by involving many stakeholders from the outset. Building on this experience and responding to other external inputs, ICARDA developed its global strategic research plan for 2007–2016. This was done through extensive stakeholder consultation throughout the non-tropical dry area countries' national authorities, research and

* This chapter draws heavily on: Belaid *et al.*, 2003.

extension agencies, universities, international and regional bodies, non-governmental organizations (NGOs), donors and other partners.

Keywords: Priority setting, priority assessment, dry areas, poverty impacts, scoring

Introduction

Research organizations apply priority assessment to outline research options and to plan for long-term investments in human and financial resources. Secondary benefits of research priority assessment include learning and consensus building among stakeholders and increased credibility for the research programme. The main output of a priority assessment exercise is a ranking of research programmes or research themes within a programme. According to Mills (1998), priority assessment efforts feed directly into planning processes, which in turn feed into processes of budgeting and resource allocation. Combined with information on current resource allocations, priority assessment provides feedback to organizations on the congruence between priorities and existing resource allocations.

The International Center for Agricultural Research in the Dry Areas (ICARDA) has a clear set of research objectives that translates its mandate into concrete goals for targeted groups. ICARDA's institutional objectives are derived from those of the Consultative Group on International Agricultural Research (CGIAR), and include poverty alleviation, food security enhancement and environmental protection. These objectives guide the process of identifying appropriate criteria for evaluating alternative research themes. Well-defined criteria that are logically linked to research objectives can help lead to credible measures of potential contributions of research themes. The most commonly used criteria for agricultural research priority assessment relate to efficiency, equity, sustainability, food self-sufficiency, food security, foreign exchange and public goods (Belaid *et al*., 2003).

Previous experiences (as documented in this volume) have demonstrated that no single priority assessment method is appropriate for all situations. The most common methods for systematic agricultural research priority assessment include congruency analysis, benefit/cost and economic surplus analysis, and scoring. To foster participation and consensus among various stakeholders, the selected procedure must be transparent. In light of the difficulty in applying an economic approach in multi-commodity, multidisciplinary, multi-objective diverse research portfolio, a simple scoring approach was used in the ICARDA regional priority assessment exercise.

The overall goal of the regional priority assessment undertaken by ICARDA was to enhance the effectiveness of the institute in working with the NARS in the region to better address their agricultural research problems and challenges posed by increasing natural resource degradation. The degradation is reflected in desertification, increased water scarcity, loss of biodiversity, and unacceptable levels of rural poverty and food insecurity. This innovative regional undertaking was based on the views of many concerned stakeholders, in particular the CGIAR, the Subregional Organizations (SROs), Global Forum on Agricultural Research (GFAR) and the NARS, to improve the economic and social well-being of Central and West Asia and North Africa's (CWANA) rural communities through enhanced productivity and sustainability of the region's agricultural production systems.

Major challenges in the dry areas

ICARDA's geographic mandate focuses on the non-tropical dry areas (henceforth referred to as dry areas) of the developing world, encompassing North and sub-Saharan Africa; Central, West and South Asia and China; and Latin America. These dry areas account for about 3 billion ha (19% of global land area) and are home to more than 1.7 billion people or 25% of the global population. About 41% of the population in dry areas depends on agriculture as the major source of its livelihood, but this figure could be as high as 81% for Ethiopia, 77% for Eritrea, 70% for Somalia and 66% for Afghanistan (ICARDA, 2007).

Characterized by water scarcity, the dry areas are also challenged by high population growth, insufficient rainfall, climatic unreliability, land degradation and desertification, frequent droughts, extremes of temperature and widespread poverty. Many farmers and pastoralists have only limited access to agricultural inputs and market opportunities. While there is a wide diversity of agroecologies in dry areas, wheat and barley represent the main components of rain-fed cropping systems, although such crops as sorghum, especially in Sudan, and cotton in Egypt and Syria (under irrigation) are also important. Faba bean, chickpea and lentil are important food legumes. Other crops such as potatoes, summer crops, oilseeds and sugarbeet are also important, especially where irrigation is available. Fruit and vegetable crops are an integral part of the farming systems and include olive, almond, fig, pistachio, apple, apricot, peach, hazelnut, grape, quince, date palm, cucumber, melon and others. Livestock, particularly sheep and goats; medicago and vetch-based pastures; and rangelands that provide feed, represent a major component of the farming system.

Food demand is outstripping production. A Food and Agriculture Organization (FAO) study revealed that food security has deteriorated in many countries in recent decades. The grain gap in 1997 was 51 million tonnes for 30 food-importing countries in CWANA. If per capita consumption of all grains remains constant at 1997 levels to the year 2025, and the UN population projections come true, and if the rate of production growth can be sustained at 2% per year to 2025, even then this deficit will easily reach 80 million tonnes by 2025 (ICARDA, 2007).

The growing food and feed deficits are reflected in poverty, hunger, malnutrition, and low income levels. Human poverty is widespread in both rural and urban communities. Currently, about 360 million people, or about 16% of the dry area population, earn less than US$1 per day. Many of the dry area countries remain in the lower half of the UN human development index, which itself correlates with the water poverty index. Economic pressures are forcing the population to extract as much as they can from the limited natural resources to meet their food, energy, housing and income needs.

The consequence of the over-exploitation of natural resources is that, except for Turkey, Kyrgyzstan and Tajikistan, the countries of the dry areas are among the most water scarce in the world, with per capita water availability continuing to decline. Furthermore, the share of agriculture in water resources is also declining, as water is diverted to other higher priority sectors. Land degradation and desertification are serious problems, both in rain-fed and irrigated areas. Much of the land degradation is attributed to human activity. Abiotic stresses (particularly drought but also terminal heat, salinity and cold/frost) and biotic stresses (insect pests, diseases and weed infestations) further compound these problems. The productivity of rangelands, which dominate the dry areas, has declined because of widespread overgrazing. Small-ruminant husbandry, central to farmers' livelihoods, is witnessing a downward spiral of overstocking, leading to land degradation and biodiversity loss, thus further aggravating poverty.

Assessing Agricultural Research Priorities for CWANA

A key factor that influenced the priority assessment process at ICARDA was the CGIAR's decision to allocate regional responsibility to its Center as described in plank 4, the *Regional Approach to Research*, which says that 'the CGIAR should adopt a regional approach to research planning and implementation in order to address the heterogeneous nature of the causes of poverty and food insecurity in different regions and integrate regional priorities with global priorities in international agricultural research' (TAC, 2000). This approach is bottom-up in order to integrate Center activities with NARS regional agricultural research priorities. It is intended to promote a multi-stakeholder dialogue, complement the ongoing CGIAR global and eco-regional approaches and ensure that identified NARS research priorities are input into CGIAR strategic planning.

ICARDA closely collaborated with the SROs – the Association of Agricultural Research Institutes in the Near East and North Africa (AARINENA) and the Central Asia and Caucasus NARS Forum (CAC-NARS Forum) – in a series of consultations that revisited agricultural research priorities in the region to enhance CGIAR–NARS integration. This approach of consultation with NARS and other stakeholders was followed by ICARDA in developing its strategic plan for 2007–2016.

The adoption of a regional approach to agricultural research priority assessment is not new to ICARDA. Over the past 3 decades, ICARDA has been actively involved in several regional meetings aimed at assessing regional priorities. Examples of such efforts include the following:

- Technical Consultation on Agricultural Research in the Near East and North Africa organized by FAO, ICARDA and International Service for National Agricultural Research (ISNAR) and held in Cyprus, 1983;

- Research Priorities in WANA and Modalities of NARS-ICARDA Interaction (ICARDA Special Task Force/Questionnaire to all Concerned Stakeholders in WANA), 1992;
- West Asia and North Africa Regional Forum: Towards a Shared Vision on Agricultural Research in WANA, organized by AARINENA, International Fund for Agricultural Development (IFAD), FAO and CGIAR, in 1995, at ICARDA;
- Central Asia and Caucasus: Seed Production and Agricultural Research Priority Setting: organized by the German Agency for Technical Cooperation (GTZ) and ICARDA (involving International Maize and Wheat Improvement Center (CIMMYT), ICARDA, International Plant Genetic Resources Institute (IPGRI) and ISNAR), in Tashkent, Uzbekistan, December 1995;
- Meeting of the CGIAR Task Force on the involvement of the CGIAR in the Central Asia and the Caucasus Tashkent, Uzbekistan, 1996;
- AARINENA Towards 2000 and Beyond: A Strategy for the Future, approved in AARINENA General Assembly held in Iran, May 1998.

As priority assessment is a dynamic process, and in view of the substantial developments that have occurred in the CWANA region over the past decade, the two subregional fora (AARINENA and CAC-NARS Forum) have recently engaged in revisiting their long-term vision and strategies with the aim of enhancing the process of regionalization of agricultural research and fostering CGIAR–NARS collaboration and integration. The expected advantages of a regional priority assessment process are several:

- **Economies of scale:** The magnitude of the challenges facing the countries of the region calls for a research agenda that would necessitate high investment (large fixed entry costs) and expertise levels that no individual NARS of the region is able to afford on its own. Therefore, a regional approach would enable the development of a critical mass of research that could lead to substantial economies of scale.
- **Synergies and complementarities:** By and large, increased scarcity of resources at the regional level has been exacerbated by a high degree of research duplication leading to a dilution of resources. Seeking regional complementarity and synergies, based on NARS' comparative advantage, would reduce research duplication and would enhance the efficient use of scarce regional resources.
- **Enhanced exchange of information:** A regional approach would promote the exchange of information, methodologies and experiences among scientists and institutions.
- **Internalizing spillovers (spillins):** Due to the nature of agricultural research outputs, many countries, especially within a subregion, are usually able to derive significant gains (spillins) from other countries' research. The possibility to tap on such valuable research spillins has very often been used by some countries as a justification for underinvesting in their national agricultural research programme. The regional approach to agricultural research represents a logic to internalize such positive externalities.

Although the above expected advantages are recognized, the following potential disadvantages are worth mentioning as well:

- **High transaction costs and bureaucracy:** A regional approach to research may be plagued by relatively high transaction costs, especially for regional coordination and monitoring of regional projects, as well as a bureaucratic environment that significantly slows down action and hence expected impact.
- **Ownership:** The research capacity of the NARS involved in a regional research programme could be quite heterogeneous. 'Strong' NARS often tend to play a dominant role within a regional framework, which in turn could lead to inadequate sharing of ownership and to potential misunderstandings and even conflicts. Building a multi-stakeholder dialogue is a very challenging task; building a multi-stakeholder ownership is clearly even more challenging.

Priority Assessment Process in CWANA

Since its inception in 1977, ICARDA has regularly consulted with the NARS on identification of its agricultural research priorities relevant to its mandate. This collaboration and experience are fundamental in ICARDA's mobilization efforts leading to a more regionally focused strategy. This consultation process takes place at the national level through annual national coordination meetings initiated since 1977. In addition, biannual subregional coordination meetings have taken place since 1979 in WANA and since 1996 in the Central Asia and Caucasus (CAC). There are five subregional coordination meetings including West Asia, North Africa, Nile Valley and Red Sea (NVRS; now Nile Valley and sub-Saharan Africa), Arabian Peninsula and CAC. Other forms of ICARDA–NARS dialogue include numerous thematic consultation meetings, such as the WANA regional priorities for livestock research held jointly by International Livestock Research Institute (ILRI) and ICARDA in 1995, 1997 and 2008 with regional NARS participation.

The consultative process is dynamic as regional agricultural research priorities regularly need to be revisited by the various stakeholders to ensure that they retain the same focus. To facilitate this continual effort, ICARDA and the two SROs (AARINENA and CAC-NARS Forum) have launched a region-wide initiative aimed at revisiting and refocusing CWANA's research priorities through an innovative consultation mechanism that relies on a bottom-up approach and broad participation including 'non-traditional' stakeholders. The emphasis on wide inter-stakeholder dialogue to set research priorities represents a departure from the traditional forms of CGIAR–NARS consultations which are often limited to a select number of national institutes, farmers and ministry officials. Thus, the key features of this new approach to assessing regional priorities by ICARDA and its partners in CWANA are the following:

- *broad dialogue* through participation of a wide range of stakeholders (includes farmer organizations, universities, non-governmental organiza-

tions (NGOs), private sector, grass-roots organizations, investors, donors and NARS);
- *enhanced coordination* and integration of CGIAR Center' research activities undertaken in the region;
- *complementarity* in tackling the regional research agenda based on comparative advantages of NARS and CGIAR Center.

The regional priority assessment launched in CWANA lays the foundation for a renewed NARS–NARS and NARS–CGIAR partnership through the consensus identification of common agricultural research priorities that are best handled at the subregional and/or regional level. The objectives are the following:

- *Revisit research priorities* at both the subregional and regional level with the aim of factoring in global, regional and national developments that have occurred over the past few years.
- *Identify the CGIAR niche* within the NARS' regional priorities.
- *Define the CGIAR strategy* in CWANA.
- *Develop an action plan* to implement such a strategy.
- *Integrate CGIAR activities*, harmonizing and integrating CGIAR activities with CWANA research priorities, so as to identify operational modalities.

These objectives were aimed at addressing concerns highlighted in previous consultations by ensuring that the process develops through a bottom-up approach and that the consultation involves as broad a range of stakeholders as possible.

Methodology of Priority Assessment (CWANA)

The priority assessment process was set up under the leadership of the two SROs (AARINENA and CAC-NARS Forum) and the GFAR and facilitated by ICARDA and other Center around three main activities: (i) a series of subregional brainstorming meetings; (ii) a questionnaire widely distributed across the region to various stakeholders; and (iii) a final regional consultation meeting that brought together (national, regional and international) stakeholders with different backgrounds and levels of responsibility within the national agricultural research system.

Brainstorming meetings

The CWANA regional priority assessment and the subsequent integrated plan were initiated through five subregional brainstorming meetings held between September 2001 and January 2002. The purpose of these brainstorming meetings was to broaden the partnership and set the stage for multi-stakeholder dialogue and discussions to identify agricultural research priorities in each subregion. Five subregional brainstorming meetings were organized for the CAC, NVRS, West Asia (WA), North Africa (NA) and the Arabian Peninsula (AP).

During the discussions and debates in these brainstorming meetings, there was an emphasis on environmental issues with calls for sustained research and policy efforts to preserve water, soil and genetic resources. The main issue discussed was water. It was suggested to address the water issue at three different levels: watershed, community and farm, and through various research themes, including water-use efficiency at the farm level, sustainable management of groundwater and surface water, safe use of waste and brackish water, salinity control and institutional and policy aspects of water use and management.

The discussions also revealed that the participants were supportive of subregional and regional research collaboration and cooperation. There was unanimous agreement that a renewed and strengthened partnership is critical for the development and implementation of research policies that would foster economic development through sustained agricultural growth while preserving the natural resource base. In these brainstorming meetings, a consensus was reached that strong research extension development linkages constitute the surest and most efficient path to agricultural development. Capacity for interstakeholder dialogue and consultation could be enhanced at all levels – national, subregional and regional – by forging alliances with 'non-traditional' partners with complementary resources and expertise to enrich the process of research priority assessment (such as universities, the private sector, NGOs, regional and national development agencies, and farmers' organizations).

Questionnaire

To gain a common understanding of the priorities of the various stakeholders, a comprehensive questionnaire was administered by the stakeholders from January to April 2002. The questionnaire, prepared by ICARDA with contributions from sister CGIAR Center and the two SROs, was designed to elicit information on agricultural research priorities, subregional and regional cooperation, and cooperation with CGIAR Center, including mechanisms to enhance integration of the CGIAR activities with regional priorities. Stakeholders were requested, through the questionnaire, to assign different weights to common criteria. Five criteria were used including productivity, poverty alleviation, sustainability, household food security and contribution to development. The agreed upon criteria were used to specify the contributions (through assigned weights) that the research alternatives under consideration are expected to make to the objectives of the region. In addition, the stakeholders were asked to assign scores to alternative research themes in order to estimate their expected contribution to the five mentioned criteria. To rank the research alternatives, the collected data were analysed to compute a 'global' scoring (based on 1–3 score: 1 = low; 2 = medium; 3 = high). The global score of each research alternative was calculated as follows:

$$Y_i = \Sigma(w_j * y_{ij}),$$

where, Y_i is the global (total) score of research alternative i; w_j is the weight of criterion j (the sum of the weights equals unity); y_{ij} is the score of research alternative i with respect to criterion j.

The survey results serve as complementary information to the subregional brainstorming meetings and the regional consultation in the process of deriving the regional research priorities. Over 500 questionnaires were distributed to representatives from National Agricultural Research Institutes (NARIs), farmers' organizations, NGOs, universities and the private sector. Although the questionnaire was sent to representatives of all relevant stakeholder categories, the bulk of respondents were from NARS representatives (researchers and managers), which has implications for the survey results. The ratio of questionnaires returned to those distributed at the subregional level was 67% (340 responses received out of 504 distributed). It would have been useful if there were more participation of grass-roots stakeholders (farmers and NGOs), as their participation is what really determines how bottom-up the process is.

Regional consultation

The purpose of the brainstorm meetings and the survey was mainly to prepare the ground for regional and broader consultations during which regional agricultural research priorities would be identified. The process culminated in a regional consultation in May 2002 at ICARDA headquarters in Aleppo, Syria. Participants (over 100) included senior representatives from 29 countries as well as representatives from the two regional organizations (AARINENA and CAC-NARS Forum), GFAR, six CGIAR Center (CIMMYT, ICARDA, World Fish Center, International Food Policy Research Institute (IFPRI), ILRI and IPGRI), two international non-CGIAR Center (Center International des Hautes Etudes Agronomiques Méditérranéennes (CIHEAM) and The International Center for Biosaline Agriculture (ICBA)), one regional Center (The Arab Center for the Studies of Arid Zones and Dry Lands (ACSAD)), one UN organization Sahara and Sahal Observatory (OSS) and a regional organization (Maghreb Union). Country representation consisted of a wide range of stakeholders including NARS, universities, NGOs, farmers and the private sector. Throughout the consultation, the emphasis was on seeking consensus on regional research priorities in order to overcome the problems confronting the region. As the conclusion of the priority assessment process, the consultation built on outcomes and recommendations of the five subregional brainstorming meetings and on the key findings of the survey.

The regional consultation meeting was structured as follows:

- plenary sessions to set the tone of the meeting, provide an overview of current agricultural research priorities in CWANA and present the Asia-Pacific Association of Agricultural Research Institutions (APAARI) experience in agricultural research priority assessment and the preliminary results of the CWANA survey undertaken in the region;
- parallel subregional working groups to revisit and finalize the agricultural research priorities identified during the subregional brainstorming meetings;

- a comprehensive analysis of subregional priorities, validated by the respective working groups, in order to identify those priorities that are common to CWANA region.

Priorities and recommendations for CWANA

This priority assessment exercise was concluded by identifying areas of common interest among the five subregions (WA, NA, NVRS, AP and CAC) to be addressed through a research agenda for CWANA. Therefore, research priorities were identified at the subregional (five subregions) and the regional (CWANA) levels. That was done through consensus among various stakeholders involved as a result of the subregional brainstorming meetings, the regional consultation meeting and the complementary information obtained from the survey.

The priorities identified can be grouped into five clusters: germplasm management, natural resources management, socio-economics and policy, cross-cutting issues, and methodologies and approaches. Germplasm management was further disaggregated into three sub-clusters: crops, animals and fisheries. Prioritization was done for research alternatives within each cluster and sub-cluster. In addition to the five clusters, three research areas were identified as major changes for the region: desertification, mountain production systems and drought mitigation. These three areas were suggested to be the focus of the CGIAR Challenge Programmes (CP), under development in 2002, to serve the CWANA region.

Germplasm improvement and biotechnology emerged as the number one priority in CWANA, within which commodity priorities were assigned to wheat, small ruminants, marine, forages, barley and cattle. In the area of natural resources management, water was given the first priority, followed by soils (as a second priority) and rangeland (as a third priority). Technology dissemination and marketing/trade were ranked as priority one and two, respectively, in the area of socio-economic and policy research. Human resource development emerged as the number one priority among cross-cutting issues. Strengthening and supporting existing subregional and regional fora was assigned the first priority among methodologies and approaches.

In order to better integrate cooperation on both the national and regional level (NARS–NARS and NARS–CGIAR), the consultations highlighted four 'gap areas' that require immediate action in order to better address the identified research priorities of the region:

- Various partnership arrangements and models are in place in the region with virtually no actual interaction among them. A thorough and critical assessment should be made of the existing partnership arrangements and corrective measures proposed to enable these partnerships to better address the identified research priorities of the region.
- Through the priority assessment process, it became clear that the two SROs (AARINENA and CAC-NARS Forum) were not sufficiently equipped with

resources to adequately address the priority needs of the region. It is therefore strongly recommended that both organizations be strengthened.

- There is insufficient involvement of many CGIAR Center such as ILRI, World Agroforestry Center, Center for International Forestry Research (CIFOR), World Fish Center, International Water Management Institute (IWMI), ISNAR and IFPRI. Considering the mandates of these Center, they could contribute significantly to improving agricultural research and development in the region.
- Some areas of development are not covered by the CGIAR Center. In particular fruit tree and dry land oil crops are gaining in importance throughout the region, but are not covered in the mandate of any CGIAR Center. It is suggested that the mandate of ICARDA be expanded to include these crops.

Key mechanisms were identified such as networks, coordination meetings and travel workshops to reinforce regional cooperation and collaboration and to facilitate implementation of regional research priorities. These interactions are essential to obtain the most from research and to ensure that research results disseminate throughout the region.

Another issue discussed in the meetings was the CPs being developed by the CGIAR. At first, the participants endorsed three key CPs of direct relevance and interest to the region, namely the CP on CAC, the CP on desertification and the CP on global genetic resources and genomics. The last of these coincides with the region's identified needs, as germplasm improvement and biotechnology were identified as two research topics of high priority for the region. Second, to ensure that regional research priorities are better reflected in the CPs, the participants strongly recommended that: (i) the dry areas be included as a special focus in the CP dealing with water and agriculture; and (ii) the CAC region, the Atlas Mountains of North Africa and the Anatolian Highlands be included in the CP dealing with mountain agriculture.

The final recommendation was to establish a committee to ensure that the region moves rapidly forward on the priorities identified and on recommendations of the meeting. It was suggested that AARINENA, the CAC Forum and ICARDA (as a facilitator) establish such a committee and be its key members.

Regional Approach in Perspective

The main purpose of a regional approach is to facilitate development that holistically fights poverty through interregional and international cooperation by involving many stakeholders from the outset. It is meant to break the hierarchical linkages in the region when assessing priorities. It is intended to elicit the views of various grass-roots stakeholders such as farmers, NGOs and private investors. While this regional process does represent a major departure from past consultations on research priority assessment, the resulting implications for decision making must be kept in perspective due to several limitations.

Limitations

One limitation in these consultative priority assessments is that the process identifies priorities but does not do so based on *ex ante* impact evaluation. This limitation is particularly an issue in a multi-objective (multi-criteria) framework because of implicit trade-offs, i.e. giving more weight to a given criterion is only possible at the expense of another. Also, the method of evaluation must be consistent, because if stakeholders use different evaluation criteria and/or assign different weights to common criteria, the identification of (shared) subregional research priorities becomes a difficult task. Thus, the priority assessment process needs to be supported by some analysis. In other words, the identification of research areas of common interest at the regional level does require analysis based on agreed-upon criteria. The criteria are used to specify the contributions that the research alternatives are expected to make towards the stated objectives of the region. In the course of this regional priority assessment process, five such criteria were used: productivity (competitiveness), sustainability, poverty alleviation (equity), household food security and contribution to development.

Another problem was the difficulty in achieving regional consensus due to the heterogeneous nature of the five subregions. The intrinsic subregional heterogeneity meant that some of the priority problems identified during brainstorming meetings did not hold across all regions. This subregional specificity generated misunderstanding during the regional consultation, as subregional representatives were concerned that priority areas specific to a given subregion would not be included in the agricultural research portfolio of CGIAR Center and their regional partners. As a result, lengthy discussions took place on the need to include some specifically subregional priorities in the regional agenda. It was reiterated that the purpose of the priority assessment exercise was to identify areas of commonality among the five subregions with regard to priority problems to be addressed through a regional research agenda.

The effective participation of all relevant stakeholder groups in assessing the regional research agenda was one of the basic tenets of this regional exercise. The valuable lessons learned from previous donor/research-driven programmes have triggered the need to develop mechanisms that would expand collaboration and dialogue through sustainable links and strategic partnerships with 'non-traditional' stakeholders. In this area, much remains to be done. At the NARS level in particular, the collaborative relationships with universities, NGOs, private sector, farmers and farmer organizations are, by and large, in an embryonic stage and need to be significantly strengthened.

To a large extent, the success or failure of subregional integration will hinge on the resource capacity of the concerned partners. Historically, capacity building at the NARS level has largely emphasized biological disciplines such as crop genetic improvement. The lack of capacity in other key disciplines, such as social sciences, combined with the shift in research focus towards relatively 'new' issues, such as poverty alleviation and natural resource management, raises a serious concern about the capacity of the NARS to adequately implement the new subregional research agenda.

Risks and threats

A regional approach to research involves various stakeholders with diverse backgrounds and interests. Achieving consensus is difficult, and there is a lack of experience with bottom-up and multi-stakeholder dialogue. A regional approach to research is about expanding collaboration and dialogue by forging strong and sustainable linkages and strategic partnerships with various stakeholders. It is also about empowering farmers' and grass-roots organizations. This shift in collaboration is a relatively novel concept in the region and will require time to be fully institutionalized.

A regional approach to research is strongly dependent on the availability and easy access to data for those in charge of implementing it. In general, data available in the region are highly imperfect and available only locally. It is not systematically made available electronically for large access through the Internet for example. The problem of lack of readily available information and relevant databases, often exacerbated by frustrating bureaucratic procedures in many countries, can affect the outcome of the approach and the sustainability of the whole process.

ICARDA Strategic Research Plan (2007–2016)

Identifying regional research priorities is a necessary but not sufficient condition to guide research for development. To be sufficient, it requires translating priorities into a concrete research agenda with appropriate funding strategies. The research agenda should be designed to address the development goals of poverty alleviation and food security improvement through generating technical, institutional and policy options that contribute to economic growth while preserving the natural resource base. Responding to this need and building on the experience in regional research priority assessment, ICARDA's global strategic plan for the 10 years from 2007 to 2016 was developed through extensive consultations with stakeholders throughout the non-tropical dry area countries, working with their national authorities, research and extension agencies and universities, as well as with international and regional bodies, NGOs, donors and other partners. Particular attention was paid to ensure that the strategy responds to the recommendations of the external programme and management review of ICARDA, conducted in 2006; the Center-commissioned external reviews conducted in 2005 and 2007; the priorities of the CGIAR system; the UN millennium development goals relating to agriculture; and the 2002 regional priorities of CWANA region.

The strategy development went through several stages over the 2-year period:

- A series of 16 externally commissioned studies were conducted in 2004/05 to identify key problems, research gaps and anticipated issues/constraints in the implementation of a new strategy. Teams of ICARDA-NARS scientists, after detailed review of past experiences, developed reports to cover five broad areas for study: socio-economic context, political and institutional context, science and technology opportunities, achieved impact and lessons learned, and environmental trends.

- Internal discussions and review of the findings, development of a first draft and early consultations with ICARDA's board of trustees (2005/2006).
- A series of consultation meetings with NARS scientists in each subregion for feedback and refinements (2006).
- Further iterations based on consultations with policy-level stakeholders (2006).
- Consideration, modification and approval by the board (2007).

Integral to the strategic visioning exercise was a review of ICARDA's research portfolio, within the context of changes in the internal and external environments, and taking into account the regional priorities of NARS and other stakeholders, the comparative advantages of the Center and its responsibility to produce international public goods.

This new strategy marks several changes within ICARDA's continuing research programmes: a broadening of the commodity focus to encompass the development of sustainable integrated production systems; an increase in emphasis on identifying pathways to poverty alleviation and income generation, enabling farmers to move from subsistence agriculture to market-oriented production and improve their livelihoods; a realization that the community institutional frameworks and enabling policy environment are vital to the successful implementation and wide adoption of new technologies; and an increased emphasis in the Center's geographic coverage of the non-tropical dry areas globally. CWANA will continue to be ICARDA's primary research platform, but with increased attention to sub-Saharan Africa, South Asia, China and Latin America (ICARDA, 2007).

Partnerships with NARS will continue to be crucial to ensure relevance and impact, and exploit complementarities in implementing ICARDA's new strategy. The Center's outreach network programmes will continue to provide a platform to address regional priorities as well as the global challenges to agriculture in dry areas. Partnerships with regional and international organizations, non-governmental organizations, civil society organizations, CGIAR Challenge and System-wide programmes and global initiatives will be strengthened and expanded.

Conclusion

Research priority assessment is a key factor in enhancing the effectiveness of public investments in agricultural science and technology. To this end, promoting R&D in developing countries requires due attention to the following:

- Revisiting research priorities to address global, regional and national developments that have occurred over the past few years, including:
    - assessing NARS capacity to address emerging research issues, such as shifting the research focus towards poverty alleviation and natural resources management to address climate change;

  - developing an action plan to implement the identified priorities, including funding and resource allocation plans that consider NARS comparative advantages, resource endowments and contributions to the development goals;
  - harmonizing and integrating the activities of regional and international organizations within identified research priorities (based on their comparative advantage).
- Understanding the evolution and complementary roles of different research partners, including NARS, advanced research institutes (ARIs) and CGIAR Center.
- Conducting *ex ante* and *ex post* research evaluation for accountability and resource allocation purposes to maximize impact and ensure relevance to poverty-targeting.

The choice of the most appropriate method for agricultural research priority assessment depends on the nature of alternatives under investigation, availability of data, analytical capacity, demand for rigour and need for consensus. Applications of congruence, benefit/cost and economic surplus methods in the literature relate largely to commodity-based, single objective and single country research programmes (Bailey *et al*., 1997).

Many priority assessment exercises utilize an estimate of the value of production and elicit the 'likelihood of effectiveness' approximated by a productivity increase (or unit cost reduction) resulting from the research, probability of obtaining the research output (a subjective assessment) and extent of technology adoption. These estimates can be easily obtained for commodity-based research, but are more difficult to obtain for non-commodity areas, such as resource-based and other cross-cutting research. Some priority assessment studies emphasize efficiency as the primary or only objective. The ICARDA research portfolio, in contrast, is aimed at achieving multiple objectives, including equity (poverty alleviation) and environmental sustainability. Incorporating multiple objectives involves an explicit or implicit trade-off between or among objectives.

Most of the studies and quantitative procedures reviewed by ICARDA scientists (Beach and Fernandez-Cornejo, 1995; Anderson, 1997; Nagy and Quddus, 1998; Muntangadura and Norton, 1999) were simplified by the fact that they referred to single country research programmes. Application of such procedures to ICARDA research themes is complicated by its international dimension (global mandate and spillovers). ICARDA's complex research agenda is characterized by a global and regional focus, multiple commodities with either global or regional mandates, non-commodity-related natural resources management research, socio-economic and policy research, and strong NARS partnerships through regional programmes, requiring a procedure that can handle each of these factors. A scoring approach is the most appropriate procedure for ICARDA's circumstances and met the immediate needs of the priority assessment exercise. As an example of such an approach, Kelley *et al.* (1995) presented impact-oriented priority assessment using an *ex ante* multi-objective (economic efficiency, equity, internationality and sustainability) framework for assessing research priorities at an international agricultural research institute.

Effective and efficient implementation of regional research priorities calls for new forms of partnerships and new roles and responsibilities for those involved. It is essential to explore opportunities to secure the necessary human and other resources to implement agreed upon regional research priorities. Assessing and implementing priorities on a regional basis, while relying on a bottom-up approach and a multi-stakeholder dialogue, will necessarily require time, resources and probably further consultations with individual NARS at the subregional level. To foster the process and ensure adequate implementation of the recommendations, a regional task force could be established under the leadership of the two SROs with participation of ICARDA. Although it is important that ownership of the process remains with the NARS and the SROs, CGIAR Center, and ICARDA in particular, will play a catalytic role in ensuring appropriate and timely implementation of the research priorities. The ICARDA strategic plan for 2007–2016 facilitates this role, as it reflects regional priorities and the underlying consultative process that was followed in identifying the research portfolio for improving livelihoods in dry areas.

References

Anderson, J.R. (1997) Policy and management work within international agricultural research. *The Australian Journal of Agricultural and Resource Economics* 41, 521–539.

Bailey, E., Aw-Hassan, A. and Tutwiler, R. (1997) *Project Formulation and In-house Priority Assessment at ICARDA: Procedure and Results*. Version 4. 23 (Unpublished).

Beach, E.D. and Fernandez-Cornejo, J. (1995) Setting research priorities in the public sector: a suggested framework for the AARC center. *The Journal of Agricultural Economics Research* 45, 3–16.

Belaid, A., Solh, M. and Mazid, A. (2003) *Setting Agricultural Research Priorities for the Central and West Asia and North Africa Region (CWANA). Toward a New NARS/NARS and CGIAR/NARS Collaboration Spirit*. ICARDA, Aleppo, Syria. vi + 48pp.

ICARDA (International Center for Agricultural Research in the Dry Areas) (2007) *Improving Livelihoods in Dry Areas. Strategic Plan 2007–2016*. ICARDA, Aleppo, Syria, x + 52pp.

Kelley, T.G., Ryan, J.G. and Patel, B.K. (1995) Applied participatory priority setting in international agricultural research: making trade-offs transparent and explicit. *Agricultural Systems* 49, 177–216.

Mills, B. (ed.) (1998) *Agricultural Research Priority Setting: Information Investments for the Improved Use of Research Resources*. ISNAR and the Kenyan Agricultural Research Institute. The Hague, The Netherlands and Nairobi, Kenya.

Muntangadura, G. and Norton, G.W. (1999) Agricultural research priority setting under multiple objectives. *Agricultural Economics* 20, 227–286.

Nagy, J.G. and Quddus, M.A. (1998) National agricultural commodity research priorities for Pakistan. *Agricultural Economics* 19, 327–340.

TAC (2000) *A Food Secure World for All: Towards a New Vision and Strategy for the CGIAR*. Technical Advisory Committee of the CGIAR Document No.: SDR/TAC: IAR/00/14.1 Rev.2. CGIAR Technical Advisory Committee Secretariat, FAO, Rome.

11 Strategic Priorities for Agricultural Development in Eastern and Central Africa: a Review of the Institutional Context and Methodological Approach for Undertaking a Quantitative, Subregional Assessment

Stanley Wood and Jock R. Anderson

Abstract

Agricultural similarities suggest that neighbouring countries in Eastern and Central Africa might gain from cooperating in key areas of agricultural research and development (R&D). This chapter describes an agricultural development strategy study conducted jointly by the Association for Strengthening Agricultural Research in Eastern and Central Africa (ASARECA) and International Food Policy Research Institute (IFPRI) intended to recognize and properly account for such regional potentials in Eastern and Central Africa (ECA) and to articulate their broad nature and scale. First, a geographic information system (GIS) was used to characterize variation in the spatial context of agriculture in ECA. Second, a process of stakeholder dialogue was initiated with a broad range of regional and national institutions and individuals including: the regional economic development authority (COMESA), agricultural research and extension agencies, non-governmental organizations (NGOs), donors, private companies, and commodity and farmer organizations. Third, specific scenarios were evaluated using two types of economic simulation models. Overall economic growth and poverty-related scenarios were analysed using a dynamic, economy-wide model (an economy-wide, multi-market model – EMM) that encompassed both sector and agricultural sub-sector components across the entire ECA region. A second set of (commodity-specific) scenarios was analysed using a single commodity, multi-region, economic surplus model to assess the economic benefits of productivity-enhancing investments in agricultural R&D.

Keywords: Agricultural development strategy, geographic information system, agricultural research priorities

The Development Context

Recent trends and the current performance of agriculture in Eastern and Central Africa (ECA) reveal a region progressively less able to meet the needs of its still – by modern international standards – rapidly growing population. With agriculture playing such a significant role in most ECA economies, slow growth in agricultural productivity has translated into sluggish overall growth and generally low per capita income levels. The region also suffers from extensive adult and child malnutrition and high child mortality (Wiesmann, 2006). The inability of national agricultural systems to respond to growing food needs has led to high levels of agricultural imports – particularly of food staples – but these appear to be only partially meeting the consumption needs of a population lacking purchasing power.

Forces such as globalization, market liberalization, privatization, urbanization, HIV/AIDS, population growth, climate change, the proliferation of powerful information and communication technologies and the changing proprietary nature of agricultural technology all contribute to redefining the landscape of challenges and opportunities facing agricultural policy makers in ECA, and thus the kinds of policy solutions required. Most of these forces have roots and expressions that extend beyond national boundaries, implying the need for broad perspectives and regional responses. Neighbouring countries might gain from cooperating in key areas of agricultural research and development (R&D). This chapter describes an agricultural development strategy study conducted jointly by the Association for Strengthening Agricultural Research in Eastern and Central Africa (ASARECA) and the International Food Policy Research Institute (IFPRI), which was motivated by a desire to recognize and properly account for such regional potentials in ECA, and to articulate their broad nature and scale (Omamo *et al*., 2007).

The Demand for New Strategic Visions and Plans

ASARECA was formally created in 1993 as an African-led, subregional organization with a mandate to variously harmonize, coordinate, support, plan and implement programmes addressing agricultural research-related issues of regional relevance in ECA. Many factors shaped the emergence of ASARECA, but the two key notions were, first, that greater effort was needed to capture likely efficiency gains to be had by taking a regional rather than a national approach to agricultural development problems faced across the region and, second, that such regional modalities might be more responsive to the real challenges faced by farmers and national research systems if they were governed by African institutions, processes and professionals. Both factors reflect some dissatisfaction with the manner in which pre-existing regional research networks and processes were seen to be captive to individual Consultative Group on International Agricultural Research (CGIAR) Center, whose research agendas were often perceived as poorly integrated or at odds with those of national systems or as being insufficiently focused on the needs of poorer farmers. Furthermore, inter-Center competition for support from CGIAR system donors, especially during periods of low growth or contraction of overall funding, has often resulted in institutional obstacles to the flow of knowledge across regional networks managed by different CGIAR Center.

The early strategic thinking and analysis that underpinned ASARECA's establishment, initial programmatic structure and research priorities were developed largely from short-term consultancy inputs coupled with regional consultations including the regional research networks, National Agricultural Research Systems (NARS; including national research organizations, universities and farmer groups), CGIAR Center and a number of donors; the US Agency for International Development (USAID) in particular. These processes placed little reliance on formal evaluation for a variety of reasons, including: the lack of – or lack of ready access to – appropriate data, the broad range of research themes to be evaluated (e.g. crops, livestock, natural resources, market development and information technology) and the limited capacity within most research networks to undertake strategic evaluation of research investment options. Strategy development relied on structured dialogue rather than access to, and interpretation of, structured evidence. Despite their shortcomings, these initial strategic planning efforts provided a solid programmatic foundation for ASARECA, including the establishment of a number of complementary regional thematic research networks under the overall coordination of an Executive Secretary, supported by a small team of planning, monitoring and evaluation specialists. Governance of ASARECA was placed in the hands of a Committee of Directors (the CD) comprising one member (the Director General of the National Agricultural Research Institute (NARI)) from each of ASARECA's ten member countries. The Committee provides policy oversight while the Executive Secretary services it and implements its decisions under the guidance of the Chairman.

One initial institutional goal of the ASARECA CD and Secretariat was to harmonize research planning and implementation into a more comprehensive and complementary set of regional networks, programmes and projects (NPPs) than it had inherited. Another was to show (primarily to the donor groups largely responsible for its creation) that ASARECA was both professional and transparent, and capable of providing the required Africa-led apex entity that could bring together diverse partners for the purposes of developing and implementing a relevant and effective regional research agenda.[1]

The approach to realigning existing networks towards a more integrated and targeted effort was to develop a new strategic framework in which the goal, purpose, strategic objectives and intermediate results (outputs) of ASARECA's efforts and investments would be openly reviewed, renegotiated and agreed upon jointly among its various stakeholders. The ambition of this approach was to develop a collective vision for ASARECA based on

[1] For example, the subregional economic community – COMESA, CGIAR Center – which had ceded control of the networks they had taken the lead in establishing, ARIs, and other NARS partners, in addition to the NARIs represented in the ASARECA CD.

a more pluralistic and consultative process than that adopted to enable its initial establishment. This realignment ran its course from 1999–2001 and resulted in the development of a new ASARECA: 'Consolidated Conceptual Framework' (CCF). IFPRI was engaged in this process through the involvement of its (then) *Agricultural Science and Technology Policy Programme*, funded by USAID's Africa Bureau, with a mandate to provide a quantitative, evidence-based orientation to the effort (e.g. through providing data on broad macroeconomic and agricultural trends, and other consistent sets of cross-country data), as well as to provide guidance on the use of economic principles to inform the assessment of research priorities, even though relatively little formal analysis was undertaken at that stage. Such principles included concern for the status of, and trends in: the scale and structure of production, consumption and trade of agricultural commodities; land and labour productivity; the costs, lag times and probabilities of success in generating new technologies; likely levels of adoption; and opportunities for the transfer (and adaptation) of existing technologies across different geographical and production environments. These principles have long been conceptualized by research policy analysts (e.g. Davis *et al.*, 1987; Anderson, 1992; Alston *et al.*, 1995) but not yet often applied, particularly at the regional level.

Once the overall ASARECA CCF had been endorsed by the CD, a separate process of reorientation was initiated for ASARECA's constituent NPPs, such that their own strategic planning and priority assessment would be properly nested within, and fully conform with, the CCF and with the strategic results for which ASARECA had agreed to hold itself accountable. This NPP priority assessment phase involved a two-year process, initiated in 2002, in which each network was charged with undertaking its own reassessment of strategic priorities using a common methodological approach established jointly by the ASARECA Secretariat and International Service for National Agricultural Research (ISNAR), with other CGIAR Center playing roles conditioned by their association with specific networks, or through providing insights into, and experiences with, specific priority assessment approaches and tools (e.g. International Livestock Research Institute's (ILRI) approach to strategic targeting and priority assessment (Randolph *et al.*, 2001), and ISNAR's early thinking on innovation systems approaches (Rosebom, 2004)). IFPRI provided commodity-specific economic analysis of the potential economic benefits of improved productivity across the ECA region (using the Dynamic Research EvAluation for Management (DREAM) model to be described in the section 'Broad Methodological Approach'), as well as more generic data on agricultural and commodity-specific trends across the region that each NPP team could include in its own priority assessment study.

The formulation of the ASARECA CCF and the subsequent harmonized NPP priority assessment process served both technical and institutional development goals. Increased dialogue and reflection helped build a new, shared sense of identity and commitment that began to enable a set of semi-autonomous and disparate regional networks to realign and to adapt to the need for more regionally and thematically coherent goals, outputs and indicators. As this process drew to a close, there was indeed a much stronger ASARECA-wide alignment of vision and focus, as individual networks completed the reshaping of their proposed outcomes and outputs to nest within the agreed CCF. This harmonization of purpose manifested itself as greater coherence among Secretariat and the NPPs at several levels, including evaluation approach, documentation templates and outreach materials. As a consequence of the rapidly evolving external and internal environment for ECA agricultural development revealed over the course of the CCF and NPP priority assessment studies, ASARECA's CD and Secretariat recognized the need to complement these studies with a review of ASARECA's strategic directions.[2] Dialogue to prepare for this study was largely conducted during 2004 between the ASARECA Secretariat, the CGIAR and donor representatives. As a consequence, IFPRI was invited to lead a study on strategic priorities for agricultural development in the ECA region, and the remainder of this chapter provides a technical description of that study. Omamo *et al.* (2007) provide the definitive description of the study and its findings and implications; our goal here is to provide a condensed summary of the study but also, with the benefit of some hindsight, to provide a more complete integrative description of the set of strategic questions that the study set out to address, the analytical steps involved and the relative strengths and weaknesses of the process and its outcomes.

Broad Methodological Approach

The regional approach agreed by ASARECA and IFPRI was explicitly strategic, but took advantage of recent developments in the richness and utility of relevant data and tools in order to undertake an analysis at high levels of spatial disaggregation (even when compared to the level of spatial disaggregation adopted by most national-scale strategic R&D studies). An overview of the approach, set in the context of the sets of strategic questions that the study sought to address, is presented in Fig. 11.1.

First, a geographic information system (GIS) was used to characterize variation in the spatial context of agriculture in ECA. This was done with high-resolution, regionally consistent spatial databases (with data typically in a 1–10 km grid format) and a common characterization framework across all ten countries in the region, thereby enabling simultaneous and comparable focus on sub-national, national and regional phenomena. Agricultural *Development Domains* (Pender *et al.*, 1999; Wood *et al.*, 1999; Omamo

[2] The likely need to review the broader strategies for reaching the agreed CCF development goals and objectives was recognized even as the NPP priority-setting studies were initiated (and was consistent with the cyclical nature of planning). But the desire to engage NPP managers and scientists sooner rather than later in a detailed examination of research activities in the light of a unified CCF drove the ASARECA management decision to launch the strategy review only after completion of the priority-setting studies.

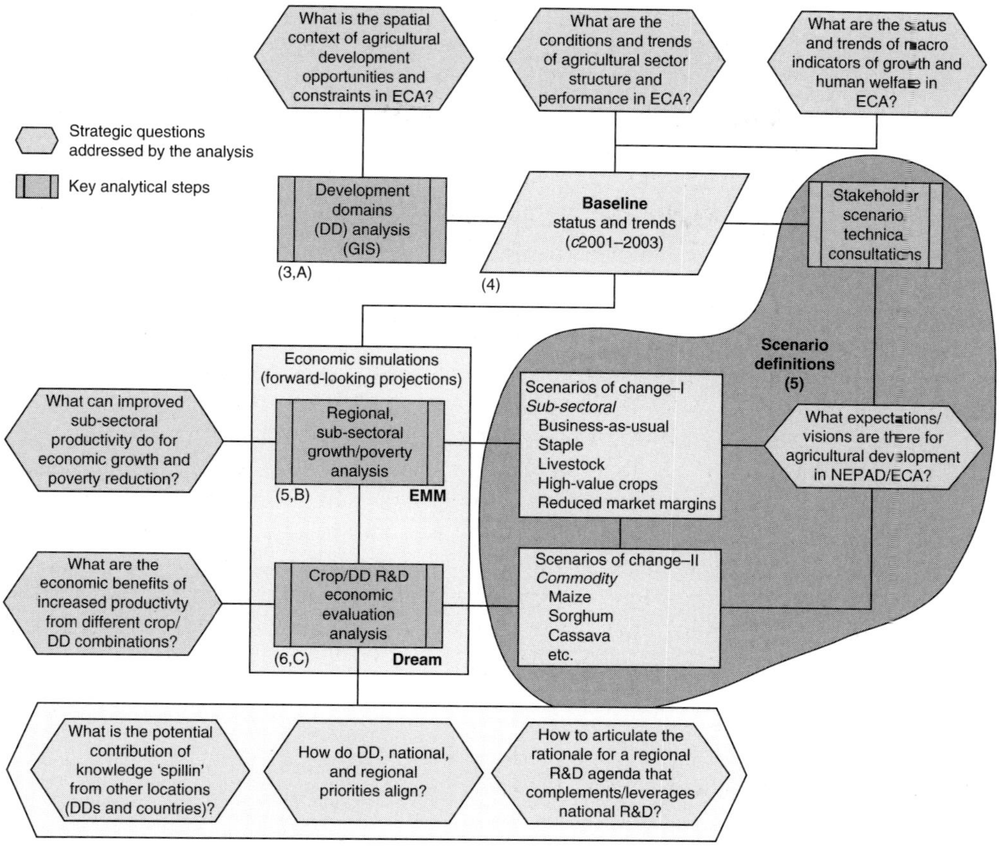

Fig. 11.1. Strategic questions and analytical components of the ASARECA study.

et al., 2007, pp. 8–9; Pender *et al.*, 2007) were delineated by combining information on agricultural potential (biophysical suitability for crop production), access to markets (time of travel to the closest town of more than 50,000 inhabitants, using road networks where they exist) and population density at each grid location (Fig. 11.2). The spatial (development) domains created by intersecting these three data layers provided the basis for comparing and contrasting spatially explicit differences in agricultural development opportunities and challenges across the region (Omamo *et al.*, 2007, Appendix A, pp. 71–84, for a complete description of data sources and methods used to delineate the eight ASARECA development domains).

Second, a process of stakeholder dialogue was initiated with a broad range of regional and national institutions and individuals including: the regional economic development authority (COMESA), agricultural research and extension agencies, non-governmental organizations (NGOs), donors, private companies, and commodity and farmer organizations (see the *Scenario Definitions* shaded portion of Fig. 11.1). These dialogues were conducted simultaneously across a number of thematic areas – including the two core themes of production

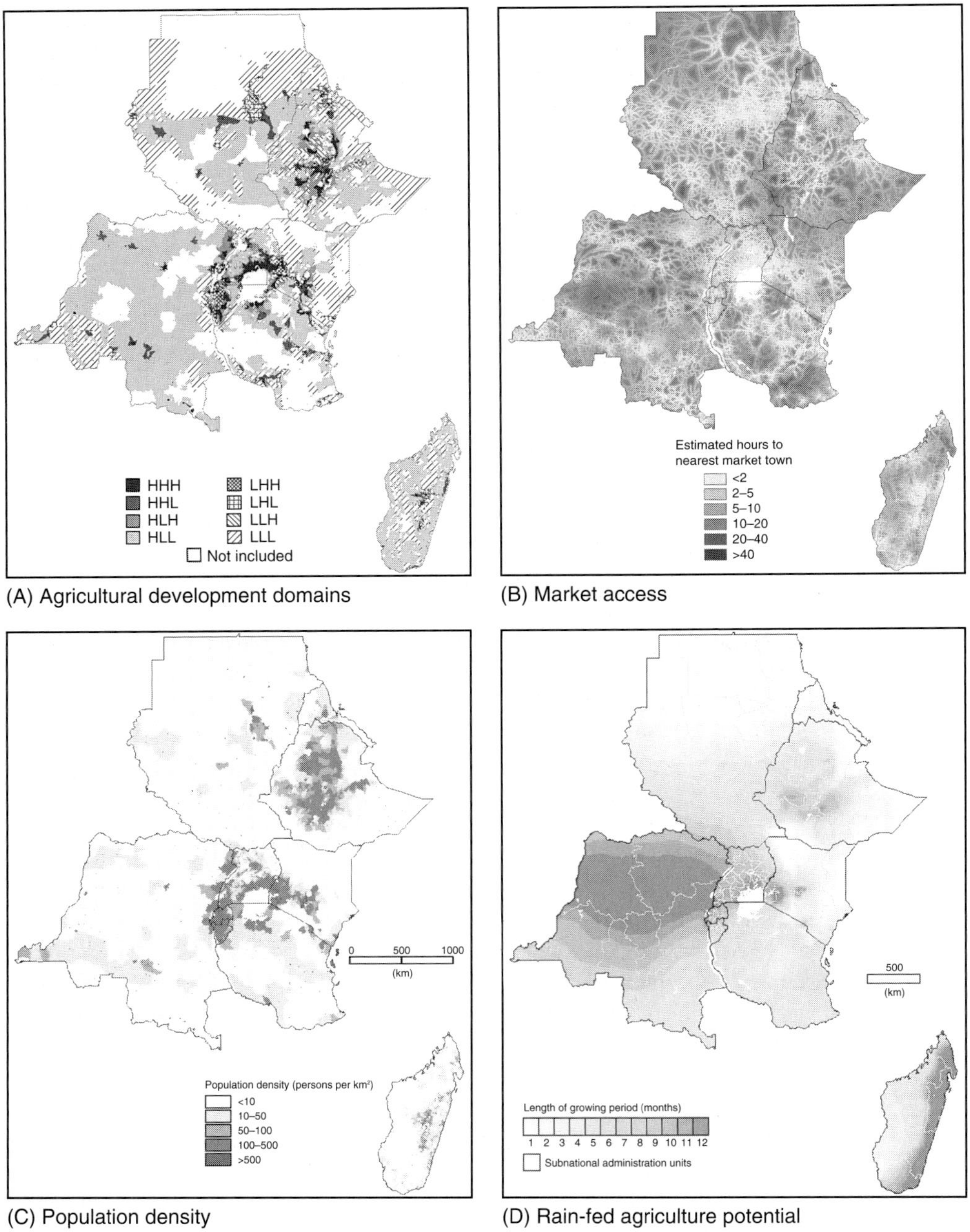

Fig. 11.2. ECA agricultural development domains, (A) defined on the basis of, (B) market access, (C) population density and (D) rain-fed agricultural potential.

systems and natural resource management. The thematic stakeholder groups were each charged with articulating future development visions and strategies for the ECA region, in dialogue with the economic modellers and analysts who would then undertake quantitative analysis. Country-specific strategic perspectives were also documented as part of this process.[3] Using this approach two broad sets of scenarios were identified as being important for evaluating the potential strategic role of ASARECA within a medium-term (3–12 year) horizon. The first set of scenarios was designed to assess: (i) the potential contribution of agriculture in promoting economic growth and poverty reduction in the region; and (ii) the comparative performance of alternate (agricultural) sub-sector growth strategies as a means of achieving target levels of agricultural growth (see *Scenarios of Change I* in Fig. 11.1). A second set of scenarios, defined at the sub-sector (commodity) scale, involved hypothesizing a range of assumptions regarding technology generation and adoption, as well as the extent to which technologies might 'spillover' from one agro-ecological zone or country to another. Such spillovers can generate additional regional economic benefits beyond those attributable solely to changes in regional trade in commodity products (see *Scenarios of Change II* in Fig. 11.1).

Third, the defined scenarios were evaluated using two types of economic simulation models. Overall economic growth and poverty-related scenarios were analysed using a dynamic, economy-wide model (an economy-wide, multi-market model – EMM) that encompassed both sector and agricultural sub-sector components across the entire ECA region. The specific units of analysis were defined on the notion of distinct domains of development opportunity, and the analysis involved explicit inclusion of quantitative measures of production, consumption, prices and trade in several important agricultural sub-sectors. The EMM model also allowed, where possible, the household income effects of agriculture-induced market changes to be estimated, by accounting for linkages between agricultural sub-sectors and other key economic sectors, as well as by the effects of trade among geographical areas. Thus, the model supports analysis at multiple geographic scales: regional, national and sub-national.

The EMM model included 33 agricultural commodities and 15 commodity groups: cereals (maize, rice, wheat, sorghum, barley, millet, oats, other cereals), root crops (potatoes, sweet potatoes, cassava, other root crops), pulses (beans, peas), oil crops (groundnuts, sesame seed, other oil crops), vegetable oil, sugar, vegetables, bananas, fruits, beverages (coffee, tea, other beverages), fibre crops (cotton), meat (bovine meat, goat and mutton meat, poultry, other meat), eggs, milk and fish.

Figure 11.3 presents some results of the growth-focused analysis. The figure illustrates the cumulative poverty reduction potential of development strategies that target specific sectors: staple crops, livestock, high-value crops, markets (reduced domestic marketing margins and trade barriers), as well as the non-agricultural sector. This was one of several sets of strategy scenarios that were examined to assess their relative effectiveness, all in terms of increased agricultural gross domestic product (AgGDP), total gross domestic product (GDP) and poverty reduction outcomes. Other scenarios examined these

[3] ASARECA strategic planning background papers are available at *http://www.asareca.org/index.php?option=publications&Itemid=8&topid=4*

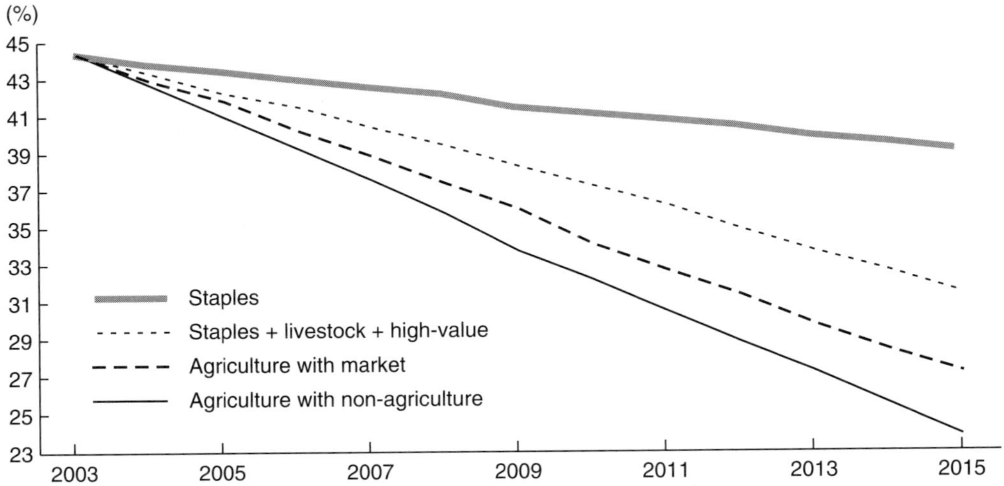

Fig. 11.3. Differential poverty reduction in Ethiopia from agricultural and non-agricultural growth and improvements in agricultural markets. (Source: Omano *et al.*, 2007. Fig. 5.6, p. 43.)

outcomes under strategies of targeting individual crops, crop groups (e.g. cereals, oil crops), product types (e.g. food staples, cash crops, livestock) and complementary sectors (marketing costs and non-agriculture rural development).

Analysis at this scale and of this type is capable of providing indicative estimates of changes in key economic and human welfare indicators of strategic importance – both to governments and other funders/donors – as a consequence of implementing a range of policies and investment strategies. Such information is of value in assessing priorities across different sub-sectors and at different points in the chain of value addition from production inputs to consumption of outputs. It can also help to calibrate expectations about the potential aggregate impacts of overall agricultural growth within a country and region and the contribution of that growth to overall economic growth and the raising of incomes, and can, furthermore, help compare and contrast the potential relative contributions of different agricultural sub-sectors (i.e. individual commodities) to overall agricultural growth. It is easy to see, therefore, a broad congruence between the outputs of this type of (EMM) analysis and consideration of the optimum configuration and balance of investments at the highest level of the ASARECA-wide agricultural R&D portfolio.

The EMM analysis revealed that 'business-as-usual' approaches to agricultural investment based on recent trends would result in agricultural and overall growth rates that were inadequate to reduce poverty in the region. In the business-as-usual scenario, no ECA country would achieve the growth rates required to meet the Millennium Development Goal (MDG) of halving poverty by 2015. On the contrary, expected growth rates would imply deepening poverty in the region. The business-as-usual outcome sheds light on the largely disappointing results in ECA of agricultural development policies in the 1980s and 1990s that concentrated primarily on reducing impediments to trade in agricultural markets (Kherallah *et al.*, 2002). Specifically, in the absence of agricultural productivity growth, both trade liberalization and reductions in domestic marketing costs resulted in GDP and AgGDP growth rates that differed only slightly from those in the business-as-usual scenario. The implication drawn was that 'letting agricultural markets work' did not equate, as had been assumed, to relegating the public sector to peripheral roles in agricultural development.

As reported in Omamo *et al.*, the EMM analysis also provided numerous insights into the nature of agricultural development that might allow countries to avoid business-as-usual outcomes. The following findings relate to Ethiopia, for instance, the ECA country for which the most complete data were available:

- Achieving GDP growth rates required to meet MDG poverty reduction targets implied threefold increases in agricultural sector and sub-sector growth rates.
- Although growth in export sub-sectors is often put forward as a pathway out of poverty for countries in ECA, the greatest reductions in poverty would come from growth in sub-sectors for which demand was highest within the region – staples, livestock products, oilseeds, and fruits and vegetables.
- Balanced growth strategies (that tap into and maximize potential synergies between sectors, e.g. raising crop and livestock productivity, and increasing crop production in tandem with postharvest processing capacity) lead to higher overall economic growth than do those featuring growth in a few sectors.
- Agricultural productivity growth alone would be insufficient to meet MDG poverty reduction targets. Growth in non-agricultural sectors and improvements in market conditions were required.
- Because poverty rates vary geographically within countries, growth strategies that take such differences into account lead to larger reductions in poverty than do those that ignore such differences.

For the second set of (commodity-specific) scenarios, a single commodity, multi-region, economic surplus model was used to assess the economic benefits of productivity-enhancing investments in agricultural R&D. The *D*ynamic *R*esearch *E*v*A*luation for *M*anagement (DREAM) model estimates the potential economic returns to local investment and regional cooperation in the generation, transfer and adoption (usually on farm, but also postharvest) of new agricultural technologies or practices (Alston *et al.*, 1995; Wood *et al.*, 2001).[4] The DREAM approach attempts to trace out a time path of both the costs and the potential benefits to producers and consumers of the agricultural products to which the technologies (practices or policies) are targeted. Furthermore, these cost and benefit trajectories are established for every geographical unit of analysis ('region').[5] This can include regions for which the technologies were

[4] The DREAM model is available for download at http://www.ifpri.org/dream.htm

[5] The ASARECA study described here did not assess likely R&D costs, so DREAM was used only to generate estimates of potential *gross* economic returns. Regions were defined at the scale of countries (ten) by development domains (eight).

specifically targeted, other regions into which the technologies are likely to spillover and those in which the technology itself may never be used but where, nevertheless, economic consequences for both producers and consumers might often be significant because of inter-regional trade.

In the DREAM analysis a set of potential 'shocks' (e.g. projected changes in on-farm productivity through investment in, and application of, new technologies) were applied to the baseline demand and supply scenarios for 20 major commodities in the ASARECA region up to 2015. Evaluating the changing space and time pattern of prices, and quantities produced, consumed and traded, with and without technical change scenarios, allows economic benefits to be computed. For the purposes of the ASARECA study, an arbitrary 1% per year increase in productivity was taken as the increment that additional R&D funding might be capable of generating. Given the baseline configuration of the distribution of production and consumption, prices, projected growth in demand, anticipated production and demand responses to price changes, among other factors, DREAM computed the potential benefits to consumers (greater levels of output at lower prices) and technology adopters (whose unit costs of production generally decline by more than the decline in market price). Using information developed on the share of production and consumption of each commodity by development domain and country (e.g. Omamo *et al*., 2007, Tables 3.8 and A.1), potential gross economic benefits were computed for each domain/commodity combination, and then scored. The priority scoring was based on the ranking of the absolute size of the benefits to producers and consumers in the ECA region together with a measure of the equality of distribution of those benefits across the region (larger benefits and more even regional distribution scored higher). One presentation of these results is reported in Table 11.1.

Table 11.1 summarizes and provides a ranking of the overall regional importance of crop and domain pairings in terms of the potential economic benefits (increased GDP) generated by an additional 1% per year increase in productivity as a consequence of increased R&D investment. Thus, raising the productivity of sorghum and millet in the high-potential, low market access, low population density (HLL) domain is projected to provide the largest regional economic benefits compared to other crop/domain pairings (the HLL domain is home to about 28% of the rural population of ASARECA and some 37% of cropland). The DREAM outputs serve to enrich those from the EMM analysis in providing commodity-specific information as well as a much more spatially disaggregated perspective by linking potential benefits to particular development domains both within countries and across the ECA region as a whole.

By linking the results of the quantitative analysis to the problem diagnoses and the scenario insights provided by stakeholder dialogues, it was possible to formulate structured, policy-focused strategic overviews of prospects and priorities for agricultural development. This was done through the generation of tables that summarized agricultural development attributes and options (Omamo *et al*., 2007, Table 3.9, p. 24) and hypothesized agricultural development strategies warranting further evaluation, for each of the eight classes of devel-

Table 11.1. Commodity-domain rankings. (Adapted from Omamo *et al.*, 2007.)

Rank	Commodity	Domain	Score
1	Sorghum and millet	HLL	54.24
2	Oilseeds	HLL	43.19
3	Cassava	HLL	41.55
4	Beef	HLL	31.41
5	Milk	HLL	28.00
6	Coffee and tea	HLL	25.54
7	Maize	HLL	23.09
8	Fruits and vegetables	HLL	20.53
9	Cassava	LLL	20.01
10	Potatoes and sweet potatoes	HLL	19.18
11	Rice	HLL	18.81
12	Bananas	HLL	17.52
13	Beef	LLL	17.22
14	Milk	HHH	17.12
15	Oilseeds	LLL	15.29
16	Milk	LLL	12.79
17	Fruits and vegetables	HHH	12.55
18	Maize	LLL	11.10
19	Sorghum and millet	LLL	10.96
20	Milk	HLH	10.90
21	Bananas	HHH	10.27
22	Fruits and vegetables	LLL	9.38
23	Potatoes and sweet potatoes	LLL	9.24
24	Fruits and vegetables	HLH	7.99
25	Rice	LLL	7.77
26	Beans and peas	HLL	7.08
27	Milk	LHH	6.63
28	Wheat and barley	HLL	6.48
29	Bananas	HLH	6.40
30	Oilseeds	HLH	6.36

Domain codes refer to those depicted in Fig. 11.2 of this chapter. Commodity scores are assessed based on the scale of contribution of domains (all commodities) and commodities (all domains) to agricultural GDP growth in the ECA region, as well as the share of regional commodity production in the specified domain (Omamo *et al.*, 2007, p. 53).

opment conditions (development domains) identified for the region. The table of priorities is reproduced in part here as Table 11.2.

The priorities of Table 11.2 were identified based on joint interpretation of the findings from the geographical, the EMM and the DREAM analyses (and their respective assumptions). Of note is that priorities and strategies that might merit further, detailed exploration can be identified for each domain. Identification and elaboration of the most complementary sets of these priorities (complementary both within and across domains) will likely offer the best promise for further articulation of the regional R&D investment

Table 11.2. Agricultural development priorities and options within ECA development domains (high agricultural potential only). (Adapted from Omamo *et al.*, 2007.)

			Example locations in ECA and potential agricultural development options	
			Population density	
Agricultural potential	Market access	Attributes/priorities	High	Low
High	High	*Greatest commercialization and diversification options* *Productivity growth* Agricultural research and extension systems Weed and pest control Soil and water management Awareness raising, consensus building on biotechnology-related opportunities and risks *Market improvement* Market intelligence (domestic, regional and international) *Linkages with non-agriculture* Storage, processing, distribution, agro-industrialization	*Example locations*: Parts of central and western Kenya, Uganda's Lake Victoria Crescent, parts of central and south-western and south-eastern highlands of Ethiopia, parts of Rwanda and Burundi *Options*: High input cereals (e.g. maize, rice, wheat) Perishable cash crops (e.g. vegetables, fruits, flowers, ornamentals) Intensive livestock (e.g. dairy, chicken, pig) Non-perishable cash crops (e.g. coffee, tea)	*Example locations*: Isolated areas scattered throughout region *Options*: As for high population density plus more extensive high-value options, e.g. cotton, tea, oil crops, fruits
	Low	*More limited technology adoption and commercialization* *Productivity growth* Agricultural research and extension systems Weed and pest control Soil and water management Awareness raising, consensus building on biotechnology-related opportunities and risks *Market improvement* Market development – i.e. infrastructure, market info systems, credit institutions, etc. *Linkages with non-agriculture* Storage, processing, distribution	*Example locations*: South-western Uganda, parts of central and western Kenya, much of the Ethiopian highlands, northern Tanzania, Rwanda and Burundi *Options*: High input cereals (e.g. maize, rice, wheat) Non-perishable cash crops	*Example locations*: Large areas of all countries: most of central DRC, southern Sudan, parts of central Uganda, Kenya and Tanzania, widely scattered areas in Ethiopia and Madagascar *Options*: Intensification in non-perishable crops – cereals, oilseeds, tea, coffee Livestock intensification; improved grazing areas

This is a partial summary of the entire priorities and options table. It does not contain results for the low agricultural potential domains.

portfolio. Thus, one virtue of planning and priority assessment approaches built around development domains is that they do not (*a priori*) bias the development focus geographically by, say, only focusing on better production environments, since the approach encompasses all land. Clearly some types of production environment could be excluded in any study, but the act of omitting a specific domain from consideration implies the need for some explicit rationalization. One such rationalization might be, for example, the greater uncertainty and higher costs associated with achieving any given level of productivity growth in more marginal development domains. To address such issues, the analysis undertaken for this study would need further extension by, for example, examining likely probabilities of R&D success in each domain. These are empirical matters, but there is significant evidence that the historic neglect of marginal lands can result in higher economic returns from agricultural development investments in such less-favoured areas compared to additional investment in better-developed, more-favoured lands (e.g. Fan and Chan-Kang, 2004).

Influence on ASARECA's Strategic Planning and Priorities

Drawing heavily on the preliminary results of the IFPRI-ASARECA study, ASARECA finalized an updated Strategic Plan (ASARECA, 2006). The ASARECA Strategic Plan (ASP) broadly accepted and adapted the situation and prioritization analyses presented in the study, and concluded by defining ASARECA's strategic approach to addressing identified priorities over the 2006–2015 period. The ASP interpreted and developed the study results in four strategic areas: priorities among commodities, priorities among domains, priorities along the production-to-consumption continuum and the potential benefits of regional collaboration.

The ASP commodity priorities were assessed according to a number of different methods in the course of the study. Table 11.3 contains a synthesis of the commodity rankings generated by each method, with a brief summary of the ranking measures and methods involved (the table includes only the top seven ranked commodities of the 15 analysed for each method). The rankings of the top 3–5 commodities appear relatively stable, although methods that favour the even distribution of benefits (by country (3) or domain (4)) give lower priority to milk, cassava and banana. Oilseeds appear most consistently highly ranked, as do beef, maize, sorghum and millet and (the very broad category of) vegetables and fruit. Another notable feature is that prioritization based only on agricultural sector impacts (methods 1 and 6) has the same four highest-ranked commodities (banana, cassava, maize, milk) but these commodities tended to move down the rankings when the GDP-based (broader, economy-wide) indicator was used (methods 2–5). These results are very congruent with the then existing commodity portfolio of ASARECA, but highlighted perhaps a lack of attention to the vegetable and fruit category (e.g. high-value, perishable cash crops). Following on from these findings, particularly building on arguments presented in the study regarding the high

Table 11.3. Comparison of commodity rankings using different methods.

		(1)		(2)		(3)		(4)		(5)			(6)
Basis		Region		Region		Region, country distribution		Region, domain distribution		Region, domain distribution, domain			Region (spillover from innovating countries)
Primary indicator		AgGDP growth		GDP growth		GDP growth		GDP growth		GDP growth			Economic surplus (welfare)
Model		EMM		EMM		EMM		EMM		EMM			DREAM
Source		*Figure 5.1*		*Figure 5.2*		*Table 6.3*		*Table 6.6*		*Table 6.7*			*Figure 6.5*
Ranking	1	Milk	1	Milk	1	Oilseeds	1	Milk	1	Sorghum and millet	HLL	1	Cassava
	2	Cassava	2	Oilseeds	2	Vegetable and fruit	2	Oilseeds	2	Oilseeds	HLL	2	Banana
	3	Maize	3	Cassava	3	Beef	3	Sorghum and millet	3	Cassava	HLL	3	Milk
	4	Banana	4	Sorghum and Millet	4	Maize	4	Cassava	4	Beef	HLL	4	Maize
	5	Vegetable and Fruit	5	Vegetable and Fruit	5	Sorghum and millet	5	Vegetable and fruit	5	Milk	HLL	5	Beef
	6	Oilseeds	6	Beef	6	Milk	6	Beef	6	Coffee and tea	HLL	6	Coffee
	7	Beef	7	Maize	7	Coffee and tea	7	Maize	7	Maize	HLL	7	Sorghum

The table summarizes results of the various commodity ranking approaches utilized in Omamo *et al.* (2007; see relevant source table or figure cited in the *Source* row for complete details). EMM – economy-wide, multi-market model. Banana includes both bananas and plantains. Methods 1 and 2 ranked the increase in ECA region AgGDP (1) and GDP (2) growth rate attributable to a 1% increase in productivity in each commodity (or commodity group). Method 3 utilizes a compound score reflecting not only the size of the regional, productivity-induced GDP growth for each commodity (same as method 2 ranking), but also a ranking of the concordance between the contribution of each crop to overall GDP growth at national versus regional levels (derived from Table 6.2). The score is simply the sum of the two rankings. Method 4 is similar to method 3, in that it uses the same overall regional GDP growth ranking, but modified by the concordance between the contribution of each crop to overall GDP growth at the domain versus at the regional level. The *product* of those two rankings is then scaled by the share of a commodity's regional production found in each domain. These commodity-domain-specific results are then summed across all domains to derive the final commodity ranking. Method 5 breaks out the individual commodity-domain rankings from method 4 (i.e. does not aggregate across domains). The second column of the method 5 ranking results indicates the specific domain of the commodity/domain combination being ranked. Method 6 draws on results from the DREAM analysis that uses a different metric of potential benefit (change in the total economic surplus derived by producers and consumers in the ECA region) and that factors in assumptions about the ability for technologies to spill over between countries/domains. The ranking is based simply on the size of the total increase in economic surplus of increasing productivity of each commodity by 1% (with Spillover). See Omamo *et al.* for further details on the individual methods and a detailed discussion of the findings.

knowledge spillover potential between many commodity improvement programmes, the eight pre-existing regional crop R&D networks were regrouped into two new consolidated programmes: one for staple crops and one for non-staple (cash) crops.

With regard to domains, the ASP prioritized four domains: HLL, LLL, HHH and HLH (see Fig. 11.2A). The HLL domain emerged from the IFPRI-ASARECA study as the one with by far the highest growth-enhancing potential in the region; accounting for close to 30% of regional population and almost 40% of land area, cropped area, pasture area and livestock population. Large segments of all ECA countries fall in this domain and it supports the production of large shares of all commodities. Complications raised by potential growth-equity trade-offs in this domain were considered to be minimal. This domain's low population density also suggests relatively low pressure on natural resources, and thus potentially fewer trade-offs between growth and sustainability. Complementary investments in infrastructure, security and market access were identified as key to leveraging the potential productivity and growth gains.

Among the three other prioritized domains, the relatively high-potential GDP gains estimated for the LLL domain reflected the prominence of livestock. Milk and beef ranked high among commodities in this respect, as did cassava and maize, both of which are important in this domain. The HHH and HLH zones produce relatively large gains because of their importance as suppliers of milk, poultry, bananas, fruits and vegetables, wheat, barley and legumes. Increased agriculture-focused investment in the LLL, HHH and HLH domains raised the possibility of trade-offs between growth and sustainability. In HHH domains, improved intensification and management-intensive techniques were identified as a particular priority. In LLL zones, concerns were noted about the fragile and uncertain environments; in HHH and HLH areas, concerns arose with respect to high population pressure. In all cases, further detailed research and analysis targeted to addressing domain-specific challenges and opportunities were outlined.

Following the publication of the ASP (ASARECA, 2006), and the subsequent consolidation of the pre-existing NPPs into seven new programmes as outlined in that strategy, each new programme has been developing an initial SP (e.g. *Strategy for the Staple Crops Programme* (ASARECA, 2008a) and the *Strategy for the Policy Analysis and Advocacy Programme* (ASARECA, 2008b)). These plans continue to place consistent and significant weight on the findings of the IFPRI-ASARECA study and the consequent ASP as the common foundation for their own strategy development. It is too early to document the extent to which these changes at the ASARECA level have had repercussions on the design and conduct of national programmes, but there is much anecdotal evidence to support the potential of such influence, e.g. the ongoing seed policy reforms of member countries, and the move towards regional harmonization of national biosafety regulations, that are both strongly influenced by ASARECA-led action (Waithaka, personal communication, 2008).

Reflections and Conclusions

The ten-country strategic study described here, including project design, stakeholder and technical consultations, data preparation and analysis, was completed in about 16 months. The study process and findings appeared to be valuable from a number of perspectives. We briefly review several aspects of the study: institutional, data and analytical tools, and subsequent implementation and influence.

From an institutional perspective, several factors facilitated the planning, implementation and satisfactory conclusion of the study: management commitment by the ASARECA Executive Secretary and CD as well as by the donor (USAID), including the proactive role of Executive Secretary as a strong 'champion' of the study; the gradual development of personal and professional trust between the Secretariat and IFPRI researchers working to support strategic policy and technology activities over the prior three to four years; and an extended period of stable staffing within the Secretariat. Acceptance of the relevance of the study was also high among management of the individual NPPs, who saw it largely as a logical progression of the drive to bring more integrated, systematic and evidence-based approaches to ASARECA planning and priority assessment. The general philosophy of movement towards evidence-based planning was also mutually reinforced by ongoing, parallel and well-publicized efforts within New Partnership for Africa's Development's (NEPAD) Comprehensive Africa Agriculture Development Programme (CAADP).

A second set of factors strongly enabling progress of the ASARECA study was the relatively good access to appropriate data facilitated by the strong linkages between IFPRI researchers and the socio-economic research community (both in and beyond the ASARECA region) coupled with the leverage of ASARECA management in facilitating access to NARS and CGIAR Center data sets, as well as reliance on pre-existing analytical approaches and tools. The strategic nature of this '*what if*?' scenario study maximized the utility of national-scale economic and agricultural sector data, and minimized the need for scientific data on the potential of specific technologies and practices. The richness of the study was undoubtedly aided by the availability and resolution of regional spatial data (especially for spatially explicit, biophysical, land-use, demographic and infrastructure variables) that proved to be highly compatible with this type of strategic evaluation, and that permitted new degrees of geographical specificity in hypothesizing development challenges and opportunities. Of particular importance was the newly developed ability to estimate the distribution and performance of crop production at a relatively high resolution, a 10 km grid across the ECA region (You and Wood, 2006). This enabled a more spatially explicit alignment of agricultural production with agro-ecological zones, population, market access and natural resource endowments to be factored into the analysis. Indeed, the application of the development domain approach relies on the availability of such spatial data which, because of their relatively high resolution (compared to second- or even third-level administrative units), provides a rich spatial picture of the regional heterogeneity of development conditions. Examined together with Table 11.2, the maps provide fresh insights into agricultural development challenges and opportunities in the region. Differences in the degree of spatial fragmentation of development conditions, for example, provide an indication of the likely type and scale of difficulties to be encountered in planning and implementation.

In practice, the domain-structured prioritization table has proved to be a successful dialogue and information outreach tool with policy makers and analysts concerned with fostering regional development perspectives and opportunities (as a complement to, or a substitute for, such summaries on a national basis). Furthermore, the utility of the table is greatly enhanced when presented in tandem with the spatial (mapped) representation of the regional patterns of development domains. One likely reason for this is that juxtaposing both the tabular and mapped information allows users to identify the precise location of specific domains that, in turn, allows even non-specialists to subjectively better calibrate/validate findings with regard to the domain-specific problem diagnoses, opportunities and priorities presented in the tables.

While the economic models used in this study have been applied and validated fairly extensively, their integration with the development domain approach was an innovative step. That is not to say, however, that there are no significant shortcomings with the analytical aspects of the study. Perhaps chief among these is the lack of any indication of the likely feasibility and cost of the specific technological (or other) interventions that might generate the hypothesized productivity enhancements. Both the EMM and DREAM scenarios were based on consistent, but arbitrary, assumptions about likely expectations for R&D-induced, sub-sectoral and commodity-specific growth. But it is not yet known in practice whether such growth targets are feasible, over what time frame and at what cost. Further elaboration of some of the strategic directions pointed to by the study, e.g. targeting of staple crops, such as millet and sorghum in HLL domains, would need to account explicitly for the expected time lags, probabilities of scientific success and user adoption, as well as the expected research, development and dissemination costs of specific interventions.

For example, it is argued that annual gains from R&D investments resulting from an assumed 1% increase in productivity across key staple crops in the ECA region would total to some US$36–56 million in gross terms. But the real economic attractiveness of such potential benefits still requires greater articulation and analysis, particularly with regard to the assessment of research-opportunity realities that the IFPRI-ASARECA study did not delve into. Specifically, two additional sets of data are required that, by their nature, will require substantial effort to forge. First, an assessment needs to be made, likely based on the subjective knowledge of relevant experts, of what gains can realistically be achieved, given the vector of current agricultural productivities and regional capacities. These might be thought of as a set of multipliers, some of which may be well above a baseline value, but others surely rather less, over some defined period of research achievement. Once this is done, the likely adjusted gross returns can be assessed, and more importantly, a subsequent reappraisal and refinement of strategies and priorities undertaken This must involve a due-diligence estimation of the likely R&D costs involved in achieving the a priori best-bet productivity gains. Such a cycle of refining the reliability of the economic evaluation of R&D impact (i.e. scoping out themes and domains of higher-potential pay-offs and only then conducting more complete and in-depth analyses within that pre-screened R&D landscape) will better allow for the impacts envisaged by the (revised) regional R&D investment programme to be compared with other public investments targeted to advancing rural economies within the region. R&D funders and managers must also grapple with the human resource dimensions of implementing any proposed research agenda, a critical aspect not addressed in this chapter, but one that we believe requires urgent attention by those concerned with achieving the needed productivity growth in the region.

Finally, there are some possible fallacies of composition in the work done to date; even if the ECA region can usefully be analysed as an 'island' with distant borders, how does one properly capture the important knowledge spillin and spillout possibilities with the many countries that in reality align closely with this island? From a comprehensive review of the literature, Alston (2002) suggests that up to one-half of the potential benefits attributable to agricultural R&D can be traced to innovation spillovers. So if, as seems very reasonable to assume, there will be major spillins to the ASARECA region from R&D and other development work elsewhere in Africa and beyond, how should the ASARECA research for development Strategy be better tailored to include efforts explicitly targeted to capture such benefits? This may be an area for more explicit attention in the future.

References

Alston, J.M. (2002) Spillovers. *Australian Journal of Agricultural and Resource Economics* 46(3), 315–346.

Alston, J.M., Norton, G.W. and Pardey, P.G. (1995) *Science Under Scarcity: Principles and Practice for Agricultural Research Evaluation and Priority Setting*. CAB International, Wallingford, UK.

Anderson, J.R. (1992) Research priority-setting in agriculture: problems in *ex ante* analysis. In: Lee, D.R., Kearl, S. and Uphoff, N. (eds) *Assessing the Impact of International Agricultural Research for Sustainable Development*. Proceedings from a Symposium at Cornell University. CIIFAD, Ithaca, New York, pp. 131–134.

ASARECA (2006) *ASARECA's Strategic Plan 2006–2015: Agricultural Research-for-Development in Eastern and Central Africa*. Entebbe. Available at: http://www.asareca.org/index.php?option=publications&Itemid=8&topid=3

ASARECA (2008a) *Strategy for the Staple Crops Programme 2008–2017*. Entebbe. Available at: http://www.asareca.org/index.php?option=publications&Itemid=8&topid=0

ASARECA (2008b) *Policy Analysis and Advocacy Programme (PAAP) Strategic Plan: 2008–2013*. Entebbe. Available at: http://www.asareca.org/index.php?option=publications&Ite mid=8&topid=0

Davis, J.S., Oram, P.A. and Ryan, J.G. (1987) *Assessment of Agricultural Research Priorities: An International Perspective*. IFPRI and ACIAR, Canberra.

Fan, S. and Chan-Kang, C. (2004) Returns to investment in less-favored areas in developing countries: a synthesis of evidence and implications for Africa. *Food Policy, Elsevier* 29(4), 431–444.

Kherallah, M., Delgado, C., Gabre-Madhin, E., Minot, N. and Johnson, M. (2002) *Reforming Agricultural Markets in Africa*. John Hopkins University Press, Baltimore, Maryland.

Omamo, S.W., Diao, X., Wood, S., Chamberlin, J., You, L., Benin, S., Wood-Sichra, U. and Tatwangire, A. (2007) *Strategic Priorities for Agricultural Development in Eastern and Central Africa*. Research Report 150. International Food Policy Research Institute (IFPRI), Massachusetts.

Pender, J., Place, F. and Ehui, S. (1999) *Strategies for Sustainable Agricultural Development in the East African Highlands*. Environment and Production Technology Division Discussion Paper No. 41. International Food Policy Research Institute, Massachusetts.

Pender, J., Place, F. and Ehui, S. (2007) *Strategies for Sustainable Agricultural Development in the East African Highlands*. International Food Policy Research Institute, Massachusetts.

Randolph, T.F., Kristjanson, P.M., Omamo, W.S., Odero, A.N., Thornton, P.K., Reid, R.S., Robinson, T. and Ryan, J.G. (2001) A framework for priority setting in international livestock research. *Research Evaluation* 10(3), 142–160.

Rosebom, J. (2004) *Adopting an Agricultural Innovation System Perspective: Implication for ASARECA'S Strategy*. ASARECA strategic planning paper No. 7. Entebbe. 14 pages.

Wiesmann, D. (2006) *2006 Global Hunger Index: A Basis for Cross-country Comparisons*. (Issue Brief 47) International Food Policy Research Institute (IFPRI), Massachusetts.

Wood, S., Sebastian, K., Nachtergaele, F., Nielsen, D. and Dai, A. (1999) *Spatial Aspects of the Design and Targeting of Agricultural Development Strategies*. Environment and Production Technology Division Discussion Paper 44. International Food Policy Research Institute, Massachusetts.

Wood, S., You, L. and Baitx, W. (2001) *Dynamic Research Evaluation for Management (DREAM). Version 3 Software and User manual*. International Food Policy Research Institute, Massachusetts.

You, L. and Wood, S. (2006) An entropy approach to spatial disaggregation of agricultural production. *Agricultural Systems* 90, 329–347. Available at: http://dx.doi.org/10.1016/j.agsy.2006.01.008.

12 Methods and Approach to Identify the Consultative Group on International Agricultural Research (CGIAR) System Priorities for Research

PETER GARDINER*

Abstract

The Consultative Group on International Agricultural Research (CGIAR) Science Council (SC) developed a new process for identifying the CGIAR System Priorities for Research (SPs) endorsed in 2005. The aims were to: (i) develop a more cohesive and focused research programme across the system of Centers and programmes; (ii) avoid dispersion of effort; (iii) mobilize capacity across the system; (iv) enhance opportunities for cooperation and coordination; and (v) define the scope of priority research for better accountability.

The new process was based on discussion and emerging consensus, rather than quantitative evaluation, with a multi-pronged approach of information gathering and analysis coupled with broad consultation with stakeholders. Priorities were developed through: (i) *deductive approaches* including analytical papers on new challenges and opportunities; (ii) *historical approaches* including review of the current and evolving research portfolios of CGIAR Center and other selected research institutions; and (iii) *inductive approaches* – identifying the demand for new research by stakeholders, and of opportunities for the potential supply of research determined by expert scientists. Three criteria were provided to guide the final discussions: (i) expected impact of the research on the major CGIAR goals (poverty alleviation, food security and nutrition, and sustainable management of natural resources); (ii) production of international public goods; and (iii) alternative sources of supply and CGIAR comparative advantage to conduct the research. The establishment of the SPs has been very important in describing what the CGIAR intends to do at the 'System level', and for providing a frame within which further planning for research implementation and resource allocation can take place. The chapter discusses the pros and cons of this consultative approach to priority identification.

* This chapter on the process used to establish the CGIAR system priorities for research was prepared by the author largely on the basis of published sources, with updates on subsequent responses and progress, for the CGIAR Science Council Secretariat, Rome, Italy, 29 September 2008.

Keywords: CGIAR System Priorities for research, CGIAR Science Council, consultative priority assessment, priority setting

Introduction

One of the responsibilities of the Science Council (SC; the body that provides technical advice to the CGIAR donors) is to review and advise the donors on the priorities and strategies for research at the CGIAR 'System level' (across all of the International Agricultural Research Center), at approximately five-year intervals. The most recent priority identification exercise was started by the interim SC at the end of 2002. It led to the development of the new CGIAR System Priorities for Research (SPs) given in Appendix 12.1. In the past, overall priorities for conducting research in the CGIAR were set (by the Technical Advisory Committee (TAC), the forerunner of the SC) based on decisions on relative priorities and resource allocation among commodities, facilitated by a modified congruence analysis. The modified congruence approach was based on the value of production-adding qualitative scoring for additional criteria, including poverty (CGIAR, 1996). This former means of priority assessment is now considered of limited value because of: (i) the complexity of goals; (ii) the 'value of production' criterion which does not work for non-market considerations such as germplasm conservation, or assistance to National Agricultural Research Systems (NARS); (iii) its inability to encompass new scientific opportunities; and (iv) the changes in funding modalities. The SC's aims were to find an alternative approach to priority identification at the global level in order to: (i) develop a more cohesive and focused research programme; (ii) avoid dispersion of effort; (iii) mobilize capacity across the system; (iv) enhance opportunities for cooperation and coordination; and (v) define the scope of priority research for better accountability.

Description of Approach

The SC took the leadership in initiating a process of system-level priority assessment (see CGIAR Science Council, 2004 for additional details). This consisted of a multi-pronged approach that involved information gathering and analysis coupled with broad consultation with stakeholders (including non-governmental organizations (NGOs), donors and scientists from both within the CGIAR system and other research institutions – NARS and advanced research institutes). The development of a preliminary set of SPs (and their associated sub-priorities) utilized three major approaches (Fig. 12.1):

- *Deductive approaches* included a broad analysis of new challenges and opportunities; the development of a set of criteria to achieve poverty reduction through agricultural research that can be used to screen future proposals; an updated congruence analysis to establish the future relative importance to be given in the overall budget to commodities (15 crops) and sectors (crops, livestock, forestry and fish) by regions, and the use of global

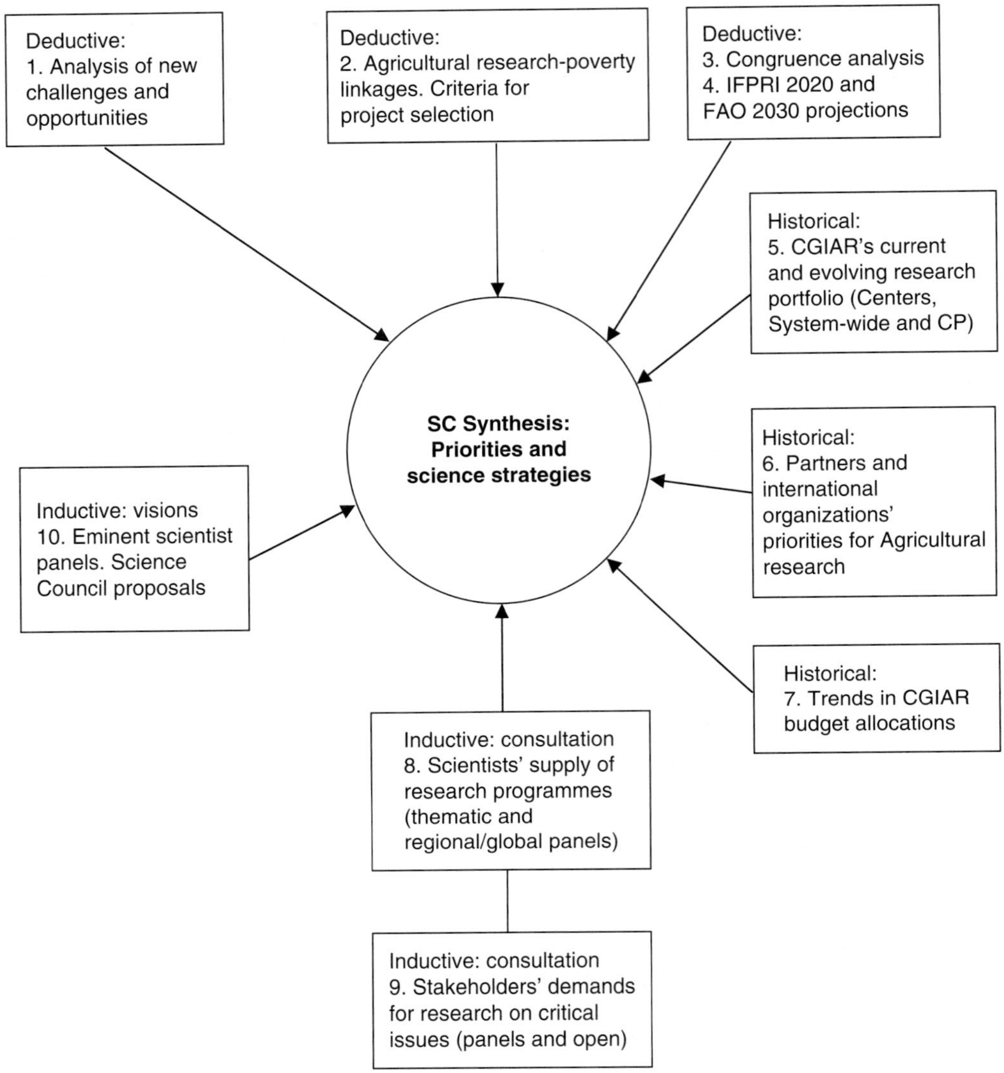

Fig. 12.1. SC's approach to identifying CGIAR priorities and strategies.

and regional projections of future supply and demand for commodities in agriculture.

- *Historical approaches* included review of the current and evolving research portfolios for: (i) Centers and the CGIAR; (ii) the research portfolios for other selected research institutions and international organizations indicative of the 'other 96%' contributing to agricultural research and development globally; and analysis of long-run trends in the CGIAR's budget allocation across outputs, crops, sectors, undertakings, regions and Centers.
- *Inductive approaches* have centred on a broadly consultative approach to identifying the demand for new research by stakeholders and the opportunities for the potential supply of research by expert scientists including members of the SC.

Step 1 identified new key challenges and opportunities in agriculture

These challenges and opportunities included: (i) trends and influences – such as private sector involvement in agricultural research; poor land- and water-use practices linked to climate change and instability; and the realization that protecting assets and incomes of the poor (e.g. through conserving indigenous livestock, maintaining access to natural resources) may be as important to their food security as providing them with new productivity-enhancing technologies; (ii) new scientific opportunities – such as in genomics and information technologies – were considered from the perspectives of their possible contribution to future research as well as the CGIAR's comparative advantage in such areas or the partnerships that would have to be forged; (iii) globalization of markets, with impacts on growth and income distribution; national economies face choices in responding to new economic opportunities as there are increasing trade-offs to be made with competing social (e.g. health, equity) and environmental goals (e.g. the need to protect environmental services); and (iv) potential and challenges of the 'New Agriculture' – which includes the identification of the means through which developing countries' needs will be met for them to take advantage of future agricultural scenarios. This includes, for example, balancing needs for the production of traditional staples with diversification towards high-value products such as fruit, vegetables, fish, livestock products, horticulture, medicinal plants, etc. Diversification also brings discussion of the relatively greater focus to be placed on post-harvest improvements, product value addition, market-oriented adjustments, food safety and increased competitiveness.

Step 2 considered world poverty concerns

This included reviews of the way agricultural research could affect poverty, taking advantage of existing literature and commissioned studies. For example, a review of the poverty reduction effects of agricultural research (Hazell and Haddad, 2001) was used to frame an understanding of the different routes through which agricultural research could be expected to have impacts on poverty. A paper by de Janvrey and Sadoulet (2002) on the indirect and direct effects of agricultural policies on poverty also enhanced the critical review of how some of the areas for research proposed early in the consultative process should be judged for their potential contributions in relation to poverty and development.

Steps 3 and 4 detailed world food production

For CGIAR commodities, a congruence analysis was conducted (in 2003) by the same method employed previously (by TAC in 1997) with values of production averaged for the 1991–2001 period. This was useful for examining swings in investments in research across a suite of commodities, as well as changes in regional allocations of effort in different sectors in the intervening period. For instance, the analysis drew attention to continuing underinvestment in livestock and fisheries and possible over-investment in pulses. These were taken as general indicators only. Two major sources were utilized: commodity projections made by International Food Policy Research Institute (IFPRI) for access to food and alternative sources of supply, and the Food and Agriculture Organization's (FAO) production projections towards 2030. These projections stressed the dual challenge of meeting increasing demands for staple foods (that can only partially be met through trade) *and* for diversification to enhance the routes out of poverty through high-value crops, animal products and fish. These two distinct goals provide a further challenge for research priority assessment to address. However, they helped demarcate the 'high-value commodities approach' which was subsequently embodied in Priority area 3 from the CGIAR's historical approach to the improvement of staples. (It is interesting to speculate whether the spikes in staple commodity prices experienced in 2007 and 2008 might change the definitions or balance between staple (now cash?) crops and traditionally higher-valued commodities like fruit and vegetables. It is likely that most commodity prices are higher at the time of writing although the opportunities for poverty alleviation (and therefore poverty weighting of commodities in the congruence analysis) would need a contemporary review.)

Step 5 examined the current and evolving CGIAR research portfolio

A review of the 196 projects presented in the CGIAR Center medium-term plans (MTPs) for 2004–2006 provided general indications of what the CGIAR was doing and how it was investing resources. While this also bears on traditional comparative advantage, it was clear that emerging issues for the Centers were: (i) a heightened concern for knowledge management; (ii) a focus on systems of innovation among farming or resource management communities and NARS; and (iii) more explicit examination of the linkages between agriculture, health and trade. The current portfolio of research was subsequently considered – at the stage of the final expert consultation – together with the new and emerging high-priority areas identified by the consultative process. This allowed estimation of synergies and differences between the two and whether some existing areas of CGIAR research needed to be modified (e.g. research brought up to date, refocused or dropped) or there were whole new areas of research that were required to be developed. For instance, the noted linkages between agriculture and health were strongly debated throughout the process, and viewed in terms of the comparative advantage of the CGIAR Center compared with existing global initiatives. Topics such as research on micronutrients in staples and reducing the effects of mycotoxins were considered central issues for agricultural research and subsequently included in CGIAR priorities. Some other areas were considered priorities for the international health sector and not included in the SPs.

Step 6 considered the priorities of partners and international organizations

The strategy papers and recent grant or research project portfolios of a selection of development agencies and partner organizations (such as the Global

Forum on Agricultural Research (GFAR)) were reviewed.[1] For example, the development banks have strategies that are aligned with the millennium development goals and which have expertise that can be deployed in partnership to augment research progress. This led to a greater awareness of the roles of the 'other 96%' of the global agricultural research system in the CGIAR discussions and the realization that research for development efforts should inevitably be conducted through the CGIAR with a greater array of partners in future. This sharpened the debate on comparative advantage (e.g. in the case of agriculture and health referred to above) and some of the principles for forming successful private–public partnerships were laid out.[2]

Step 7 considered the historical CGIAR budget allocation

Historical resource allocations by CGIAR undertaking and outputs (as then defined) were examined and reviewed in relation to projected budgets for similar areas of effort in 2006. Such demarcations define broad types of research (i.e. productivity, conservation of biodiversity, etc., rather than research effort by commodity or sector). For instance, it became clear that there had been a steady decline in productivity-enhancing research in the CGIAR with a concomitant increase in natural resources management research. This allowed questions to be posed as to the relative balance of research in the two domains that the CGIAR should consider for the future, and the nature and scope for any renewed efforts in germplasm enhancement.

Steps 8 and 9 followed a consultative approach for CGIAR priority assessment

First, a set of some 20 position papers was written by experts on the key issues in agricultural development and research and they were shared with the other participants in the priority assessment exercise and the public at large.[3]

Constraints were then further identified and elaborated through the discussions by a series of regional panels, as well as a global panel which functioned virtually, each made up of some 20 experts with different disciplinary and organizational backgrounds. To broaden the consultation, and seek majority opinions, the list of research activities was offered to stakeholders through an open electronic forum. More than 800 participated, including significant numbers of NGO and NARS representatives.

[1] For example, see Chapter 5 in reference 3, Science Council, August 2004.

[2] See Leisinger (2003) quoted in Science Council August 2004 (CGIAR Science Council, August 2004).

[3] Papers provided updates and common background to inform the special panel discussions (on matters such as food security, fisheries, forestry, biosciences, socio-economic policy, the state of NARS and communications technology) relevant to a range of priority areas. See Footnote 3 and the Priorities and Strategies Consultation Web page where these are collected, at http://www.rimisp.org/cgiar-ps2

On the 'supply' side, thematic panels of scientists were asked to translate priorities into researchable activities. These panels proposed researchable themes (sub-activities) within each of the categories of activities identified by the earlier steps. The relative importance of priorities was then established by regions. Regional panels of scientists were asked to identify additional areas and to allocate given incremental resources to the researchable sub-activities identified by the thematic panels. The figure in Appendix 12.2 illustrates the conceptual framework used by the interim SC to help in clustering suggestions arising from the consultative process on potential new areas for research. Areas for research that had emerged from discussions up to the point were evaluated for how such work might have a potential role in contributing to development pathways and poverty alleviation. Consultations to this point had focused principally on new areas for research, potentially additional to the existing portfolio.

Final expert consultation

Draft research priorities (initially described across ten priority areas – see Box 12.1.) were developed by the SC considering the new opportunities identified and developed over the 2-year period together with the existing portfolio of CGIAR research. For this prioritization, a series of expert consultations was conducted around coherent themes informed by existing research but encompassing the new suggestions. For example, some traditional areas (for the CGIAR) such as genetic improvement of specific traits (in staple crops) were considered but only with respect to a more specific focus on drought and salinity tolerances. Some areas were quite new for the CGIAR or built on only scattered research experience, e.g. improved production and processing systems for high-value commodities.

Box 12.1 The Ten Topics for Detailed Discussion and Scientific Assessment During the CGIAR Priority Assessment

1. Conserve and characterize genetic resources
2. Genetic improvement of specific traits
3. Improved water management and use in agriculture
4. Better management and use of forests and forest landscapes
5. Better soil and land management and use
6. Improved production and processing systems for high-value commodities
7. Enhancing resource efficient and equitable forms of livestock sector growth
8. Improved management and use of aquatic resources
9. Policy and institutional innovation to reduce poverty and hunger and to enhance competitiveness of smallholders
10. Strengthened national and regional capacities for agricultural research and rural institutions

Outline descriptions of research prioritized under the ten areas were shared with the CGIAR and its stakeholders in October 2004, and the approach was widely discussed in several fora, including at the CGIAR's Annual General Meeting in 2004. During January and February 2005, the Council organized discussion meetings on each of the ten priority areas to examine and refine the rationale and possible research content for each area. Approximately 30 participants took part in each of the different meetings, including invited scientists, CGIAR Center researchers and SC and secretariat members. Some of the draft priority areas were substantially revised and consolidated on the basis of the outcomes of those meetings. The boundaries for priority research emerged more clearly from the discussions (e.g. whether it made more sense to include fisheries within a water management approach to agriculture or to see the priority for the stabilization of small-scale fisheries production as a discrete area for research although contributing to production, income and environmental goals). This sort of consideration led to specification of a number of areas (20) against the original ten) and the teasing out of general and specific goals for research in each area. Finally, the CGIAR Center Deputies for Research Committee provided comments on a subsequent draft of the outcomes of those meetings.

Criteria for identifying priorities for CGIAR research

The consultative stages of the process of priority assessment led to the identification of a number of potential topics for CGIAR research. Each topic responds to important issues in the new agricultural environment emerging with the globalization of agriculture that potential impacts from new undertakings should be determined and balanced against current undertakings of the system as a whole. In identifying the priorities for research, the SC utilized the following three criteria for reference:

1. Expected impact of the research on the major CGIAR goals (poverty alleviation, food security and nutrition, and sustainable management of natural resources), taking into account the expected probability of success and expected impact if successful;
2. Production of international public goods;
3. Alternative sources of supply and CGIAR comparative advantage in the conduct of the research.

The SC provided brief statements (or existing texts) to clarify these criteria and other concepts for all participants to the expert consultations (as detailed in CGIAR Science Council, 2005). The overall expert opinion-based evaluations of research options were intended to implicitly include whether the proposed research met these criteria in the judgement of the discussants. No further quantitative assessment of these criteria by research area was attempted at this stage. External experts were invited both to introduce the best of world science and to maintain realism in the assessments of CGIAR Center capacity in the different areas.

Methodology/data sources for criteria appraised

A substantial amount of data was accumulated (panel outputs and syntheses, think pieces, key literature, etc.) and these various influences and discussions of some of the newly proposed research areas were summarized at an intermediate step in an internal report (CGIAR Science Council, 2004).

A critique of the early consultative process has been published (as part of CGIAR Science Council, 2004). In general, it was considered a low-cost exercise and somewhat unstructured in its early stages. However, it was considered to have stimulated creative interventions with adequate time for discussion. There was a low level for buy-in through the electronic panels. Drawbacks included the unequal performance across discussion panels. However, the large amount of data generated, the emerging consensuses on needs and the growing knowledge among stakeholders of the issues and the scope of the exercise meant that people entered the expert discussions with a largely similar understanding of the major challenges facing agriculture and development. The criteria were not applied quantitatively, but they were brought to the fore to shape all of the expert discussions (including, for instance, experts on poverty studies who were given central roles in the meetings).

Establishing the process for the future (i.e. a new priority assessment exercise in five or more years) may be simpler given this earlier experience. Opportunities for better linkage of impact assessments (conducted by the SC's Standing Panel on Impact Assessment (SPIA) by Center or through consortium approaches) of research under the new portfolio could eventually feed into a new round of priority assessment.

Key findings

- The early stakeholder consultations looked at *additional* areas that the CGIAR might address, and discussions delivered lists, which tended to be a mixture of areas, topics for research and approaches, that had to be recast subsequently as researchable issues.
- A complete exercise needs to balance new and existing areas adequately – and this was done through the examination by the later scientific consultations of goals and scope of research for each priority. Articulating the scope of research also helped delineate boundaries to the ambition and between SPs.
- The process 'updated' the CGIAR research portfolio, bringing it closer in line with efforts to lead development through income-generating approaches and with the science of the private sector.
- It allowed useful debate between the various stakeholders in the CGIAR of the meanings of the various criteria and how the CGIAR should operationalize them. For example, the meanings of international public goods (IPGs), comparative and competitive advantage, output-focused planning, research for development partnerships, etc. have evolved through discussion and

are better accepted as operational concepts. (Indeed the SC subsequently developed a further elaboration of the concept of IPGs, and the required partnerships for research for development, to help guide Center thinking in the move from priority assessment to implementation (Ryan, 2006; Science Council Secretariat, 2006)).

- Capacity building was identified as a basic and cross-cutting CGIAR function and was separated (but not eliminated) from the consideration of *research* for development priorities.
- The SPs were endorsed by the CGIAR (in 2005) including general goals, specific goals and scope of research by SP – which was assumed to lay the groundwork for further research planning by groups of Centers and their partners.
- There was no attempt to rank the SPs – assuming that the CGIAR and donors preferred a portfolio approach.
- Similarly the SPs were not viewed as requiring equivalent funding to accomplish the different goals set.
- It is probable that pair-wise *ex ante* impact comparisons among priorities would be possible in the future, once their research has been defined. This comparison may be more feasible within priority areas, e.g. between different sorts of genetic enhancement research, than between priority areas, e.g. between germplasm conservation and institutional research.
- Virtual panels can be effective in providing input in a consultative process, and the wider net creates colleges of knowledgeable stakeholders for future efforts – although fatigue builds up and such major efforts at priority assessment should only be repeated at suitable intervals.

Discussion

The process was, at the time of the endorsement of SPs, thought to be in large measure successful. It led, for instance, to the express stated inclusion of important new areas for the CGIAR in research for poverty alleviation and sustainability such as on high-value fruit and vegetables, exploitation of underutilized species, research on markets, landscape-level approaches to natural resource management and the rehabilitation of degraded lands. Several areas of CGIAR research continued to figure strongly as before, particularly where the priority assessment process confirmed that the CGIAR has a role in the production of IPGs from public funds – such as improvements to staples not addressed by the private sector or in natural resources management research. However, as indicated, strategic changes were also made, for instance, within the existing approaches to staples (particularly tilting new research emphases towards abiotic stress tolerance and micronutrient enhancement, but also encouraging a more globally coordinated approach to breeding with NARS seen as key partners rather than recipients and testers of CGIAR germplasm).

The priorities were developed at a relatively high level of aggregation. While the specific goals for research and their scope had been derived and agreed upon, these goals were expected to be refined through individual SP planning debates. From the first attempts at framework plan development (for SP 1A on the conservation of staple crops) which followed quite closely the structure of specific goals and outputs proposed, it seemed likely that the process could encourage the development of global research plans as the SC intended. There are exciting opportunities to develop research in areas that the CGIAR has not tackled previously among Centers (e.g. landscape-level research) and with partner groups external to the system (such as on high-value commodities) which have yet to be implemented. The utility of the priority assessment process is considered below.

Utility of findings

The establishment of the SPs has been important, since they define overall what the CGIAR intends to do at the system level, and where it will focus the bulk of its resources. While TAC's priority assessment did so descriptively, the intention (so far unrealized) of the SPs is to move towards collaborative planning of research and outputs to develop more programmatic ways of working. These intentions are similar to the goals of the Change Management process now (in 2008) being addressed by the CGIAR system. It was intended that the SPs would provide a framework within which further discussions and plans for implementation of research could take place. This framework would provide opportunities to bring about a more cooperative global approach to research (the CGIAR with research and development partners, as well as between Centers), rather than approaches solely arising from the plans of individual Centers.

Confidence in the consultative process (as was expressed at the time of the endorsement of SPs) led to calls by the SC and some donors for: (i) developing Framework Plans to implement SPs and to meet the specific goals for new research linked to the global activities of others; (ii) developing potential vehicles for research (such as challenge programmes aimed at SP research) (iii) funding the SPs – without which any priority assessment exercise is an abstract process; and (iv) sequencing other debates (e.g. related to CGIAR structure because the substantive research is now better defined).

However, the proposed process for the translation of priorities into new research has stalled. The recent independent review of the CGIAR (CGIAR Independent Review Panel, 2008) presents it in this light: 'The Centers and Science Council have struggled to find a common vision for implementing the System Priorities. The recent decision to discontinue work on framework plans, and the mandate of the Change Design process to develop a small set of outcome-oriented strategic objectives for the system, would suggest that the CGIAR has decided that the system Priorities have not been effective.' The review continues: 'Centers resisted major restructuring of their research programmes around System Priorities for a variety of reasons.' It does not elaborate on these reasons. 'An important lesson from this exercise', the review report continues, 'is that the Centers must have a major role in developing the strategy that they will implement, to ensure both ownership and feasibility'.

The independent review states that attempting to move the priority assessment process through to implementation has brought several issues to light:

> [A] resource allocation mechanism has identified important and challenging preconditions for success. The initial framework planning exercise has demonstrated that moving to a more programmatic, cross-Center research approach may incur substantial transaction costs and challenge Center research strategies. The exercise has shown how research in different Centers might fit together in a systemwide research framework and how special inter-Center initiatives can strengthen that. It stimulated consideration of how boundaries might be put around CGIAR research. In addition, it highlighted important issues for any future consortium of Centres, such as how genetic improvement, natural resources management and policy research should be managed together across Centres.

It is to be hoped this greater understanding will be carried through into the future planning of change and will inform the adoption of a more programmatic approach by the CGIAR in the future.

Limitations/constraints

The priority assessment was a long consultative process (the process was planned at the end of 2002; Step 1 of the process started in March 2003, and the process culminated with the endorsement of priorities by the group at the annual general meeting in December 2005). However, given the large number of stakeholders with quite different perceptions, it is unlikely that attempting a more quantitative approach to strategic global issues, based on many assumptions and trade-offs, would have brought a swifter result.

Part of the length of the process was the early emphasis on defining only *additional* research areas that might be included by the CGIAR. Additional areas for research can only be accommodated assuming additional resources are available to support them. Thus, while such definition was a necessary first step, the second part of the process addressed a more comparative view of current and future research areas, and trade-offs among these areas in pursuit of focus, and determined options at various levels of funding in the future.

The process required the active involvement of a large number of people: this included the SC and SC secretariat staff; the appointment and management of discussion panels working virtually, convenors, rapporteurs and writers; assistance in the management of the process and the establishment of a Web site for sharing data (through an external organization); ten face-to-face meetings in Rome with CGIAR- and non-CGIAR-invited participants (20–30 per meeting); and discussion and broadcast of the results in several fora. An inclusive, consultative process at the system level rather than a congruence or Center-by-Center approach to priority assessment had not been carried out before. Given the overall size of the CGIAR investment, this specific attempt at priority assessment may be considered a relatively low-cost approach. However, the proposed method for further inter-Center planning would have had consequential additional costs if pursued, which would come on top of investments in priority assessment exercises by each of the Centers (described elsewhere in this book).

An advantage in defining the scientific scope of SPs within the overall goals for the CGIAR system is that CGIAR can be clear about its contribution to international efforts and can staff and resource its contribution accordingly. However, this does not mean that the individual priorities are disciplinary. First, achieving the specific goals and outputs will require many types of research approaches to be integrated in different ways *within* the individual SPs. Second, while component research has to be carried out in pursuit of a limited number of objectives, specific development goals will be addressed by combinations of research outputs. Although the SPs report (CGIAR Science Council, 2005) states how cross-cutting research and capacity-building functions are built into the portfolio, stakeholders developed different appreciations of these functions, which should be resolved. This also highlights the fact that the definition of research priorities is only part of the exercise. There are communication and joint-learning steps to be completed as well, so that follow-up is initiated and implemented with the same ethos by all parties.

References

CGIAR (1996) The CGIAR research agenda: facing the policy challenge. In: *The Report of the CGIAR Mid-Term Meeting, Jakarta, Indonesia, May 20–24 1996. Summary of Proceedings and Decisions*. Available at: http://www.cgiar.org/corecollection/docs/csop0596.pdf

CGIAR Independent Review Panel (2008) *Bringing Together the Best of Science and the Best of Development*. Independent Review of the CGIAR System, Report to the Executive Council, Washington, DC, USA.

CGIAR Science Council (August 2004) *Report on CGIAR Priorities and Strategies for the Period 2005–2010 (Outcomes of the Consultative Process)*. Science Council Secretariat, FAO, Rome, Italy, p. 208 plus 3 Appendices. Available at: http://www.sciencecouncil.cgiar.org/fileadmin/user_upload/sciencecouncil/Reports/Item_3_-_NewCGIAR_PS_REPORTJULY04_R.pdf

CGIAR Science Council (December 2005) *System Priorities for CGIAR Research 2005–2015*. Science Council Secretariat, FAO, Rome, Italy, pp. 86. Available at: http://www.sciencecouncil.cgiar.org/fileadmin/user_upload/sciencecouncil/Reports/SCPriorities_prFinal_ l-r_.pdf

CGIAR Science Council Secretariat (2006) Partners in research for development. In: *CGIAR Science Council. Positioning the CGIAR in the Global Research for Development Continuum*. Science Council Secretariat, Rome, Italy, pp. 25–44. Available at: http://www.sciencecouncil.cgiar.org/fileadmin/user_upload/sciencecouncil/Reports/REPORT_of_R4D_website.pdf

de Janvrey, A. and Sadoulet, E. (2002) World poverty and the role of agricultural policy: direct and indirect effects. *Journal of Development Studies* 38, 1–26.

Hazell, P. and Haddad, L. (2001) *Agricultural Research and Poverty Reduction*. IFPRI Discussion Paper 34. IFPRI, Washington, DC.
Ryan, J. (2006) *International Public Goods and the CGIAR Niche in the R for D Continuum: Operationalizing Concepts*. In: CGIAR Science Council. *Positioning the CGIAR in the Global Research for Development Continuum*. Science Council Secretariat, Rome, Italy, pp. 1–24. Available at: http://www.sciencecouncil.cgiar.org/fileadmin/user_upload/sciencecouncil/Reports/REPORT_of_R4D_website.pdf

Appendix 12.1

The CGIAR systemprioritiess for research

Priority area 1: Sustaining biodiversity for current and future generations
Priority 1A: Conservation and characterization of staple crops
Priority 1B: Promoting conservation and characterization of underutilized plant genetic resources to increase the income of the poor
Priority 1C: Conservation of indigenous livestock
Priority 1D: Conservation of aquatic animal genetic resources

Priority area 2: Producing more and better food at lower cost through genetic improvements
Priority 2A: Maintaining and enhancing yields and yield potential of food staples
Priority 2B: Tolerance to selected abiotic stresses
Priority 2C: Enhancing nutritional quality and safety
Priority 2D: Genetic enhancement of selected species to increase income generation by the poor

Priority area 3: Reducing rural poverty through agricultural diversification and emerging opportunities for high-value commodities and products
Priority 3A: Increasing income from fruit and vegetables
Priority 3B: Income increases from livestock
Priority 3C: Enhancing incomes through increased productivity of fisheries and aquaculture
Priority 3D: Sustainable income generation from forests and trees

Priority area 4: Poverty alleviation and sustainable management of water, land and forest resources
Priority 4A: Integrated land, water and forest management at landscape level
Priority 4B: Sustaining and managing aquatic ecosystems for food and livelihoods

Priority 4C: Improving water productivity
Priority 4D: Sustainable agro-ecological intensification in low- and high-potential environments

Priority area 5: Improving policies and facilitating innovation to support sustainable reduction of poverty and hunger
Priority 5A: Science and technology policies and institutions
Priority 5B: Making international and domestic markets work for the poor
Priority 5C: Rural institutions and their governance
Priority 5D: Improving research and development options to reduce rural poverty and vulnerability

Appendix 12.2

This appendix indicates the framework used in clustering the suggested research priorities against pathways out of poverty. The numbers in brackets relate to the 32 areas for research which had emerged from suggestions up to the point in the process (From CGIAR Science Council, August 2004).

Pathways out of poverty
CGIAR research prorities
Asset positions
Control over assets by households and communities: natural, physical, human, financial and social capital
Increase access to assets for the rural poor, especially by gender (12)
Typology of household livelihood strategies
Poor as smallholders in FALs
Irrigated lands
High quality rain-fed lands
Urban and peri-urban
Water management (1)
High value-added crops and systems, including animals and fish (5)
Increase production of staple foods, including maintenance research (9)
Vegetables and fruits production systems (17)
Poor as smallholders in MALs
Desert margins
Mountain margins
Forest margins
Coastal margins
Isolated areas
Drought and salinity resistance (3)
Agricultual systems and INRM for unfavourable and risky environments (4)
Artesanal fisheries management and coastal margins(7)
Sustainable use of rangelands (CWANA) (10)
Community forest management for marginal environments(15)
Agroforestry for food, fuel, and fodder (14)
Integrated weed (IWM) and pest (IPM) management (24)
Smallholder provision of environmental services (27)
Poor as rural landless
Agricultural workers
Non-agricultural rural workers
Labour-intensive agricultural (indirect effects) and rural development strategies (29)
Postharvest value-added, commodity chains, new agriculture (30)
Poor as consumers of food
Urban poor
Rural net buyers of food
Animal health and human food/health safety (8)
Food/water safety and more nutritious foods (16)
Reduce post harvest toxicity (26)
Poor as socially excluded groups
Unorganized smallholders
Female smallholders
Indigenous populations
Empowering agricultural research with gender focus (25)
Participatory farmer-breeder management of crop gene pools (23)
Context where assets are used
Access to markets
Marketing innovations to link farmers to national and international markets (28)
Rural institutions
Strengthening producers organizations (strategy)
Germplasm distribution systems (32)
Policies
Global trade policies and opportunities for smallholders in globalized markets (21)
Policies for sustainable NRM (20)
Policies and institutions to enhance adoption of new technologies (22)
Public goods: R&D capacity
IPR, IPG and CGIAR research. Public and private roles (11, 31)
Strengthening NARS (strategy)
Extended germplasm conservation for wild relatives, orphan crops (2)
Biosafety and gene flows (13)
Indigenous livestock genetics (6)
Forest genetic resources (19)
Adaptation to climate change (18)
Well-being
(objectives)
Poverty reduction
Vulnerability reduction, food security
Health, nutrition,
Sustainability
Social incorporation
Quality of life
Assets accumulation
Political process
Stakeholder participation
Government priorities
CGIAR and donor priorities

13 Synthesis and Options for Enhanced Priority Assessment for Agricultural and Natural Resources Research

George W. Norton and David A. Raitzer

Abstract

Lessons are synthesized from chapters in this book, and a possible way forward is suggested for meeting priority assessment challenges under the multiple objectives of Consultative Group on International Agricultural Research (CGIAR) Centers, National Agricultural Research Systems (NARS) and other research institutions. Over time, agricultural research system goals and objectives have broadened in directions that challenge previous priority assessment tools. Meanwhile, the CGIAR Centers have independently developed their priority assessment approaches over time, and substantial differences in approaches have evolved. However, a common set of challenges and good practices is evident across the various exercises, which can be best handled if explicitly considered during the design of priority assessment efforts. Adherence to a few basic identified priority assessment principles may help improve consistency in methods across institutes and programmes. So as to help expand the application of these principles, the chapter provides brief suggestions for priority assessment methods that incorporate impacts of natural resource management and policy-oriented research.

Keywords: Research priority setting, priority assessment, *ex ante* impact assessment, research evaluation, impact pathways, multiple objectives

Introduction

Priority assessment applications described in this book for CGIAR Centers, a regional association, a CGIAR Challenge Programme, and the CGIAR Science Council provide a wealth of lessons and challenges for improving priority assessment within the CGIAR system and its partners. They also highlight the need for additional methodological work, especially in the area of priority assessment for natural resource management (NRM) and policy research. The purpose of this chapter is to synthesize the ideas and lessons from the earlier chapters, and to suggest a possible way forward in meeting priority assessment challenges under the multiple objectives of the CGIAR and other research systems.

The diversity of priority assessment approaches applied in this book is striking, and while each Center or institution has a unique mandate, a greater degree of consistency in priority assessment methods may be attainable and useful. Many Centers have refined their priority assessment approaches over time, and differences in approaches are driven by history, expertise available for priority assessment and management styles in individual institutions. Adherence to a few basic priority assessment principles may help improve consistency in methods across institutes and programmes. These principles are highlighted in the discussions below. As noted in several chapters, CGIAR system and National Agricultural Research System (NARS) goals and objectives have broadened over time in directions that challenge previous priority assessment methods, although not basic priority assessment principles.

One of the objectives of the CGIAR and many NARS is to enhance sustainability of the natural resource base, which requires policy and management as well as biological research. Roughly a third of the CGIAR portfolio directly addresses NRM issues (Kelley and Gregersen, 2005), and as much as half addresses policy issues (Raitzer and Ryan, 2008). Research institutes have often struggled with this issue in applying priority assessment methods. Therefore, this chapter concludes with brief suggestions for priority assessment methods that incorporate impacts of NRM and policy-oriented research.

Lessons from Priority Assessment Studies

Several lessons emerge from priority assessment at CGIAR Centers and at other research institutions. They relate to objectives, methods, processes and resources required for research priority assessment.

Objectives

One of the first principles of priority assessment is to define the goals or objectives for the organization. Chapter 12, which describes the Science Council approach to identifying CGIAR system priorities, defines the CGIAR goals as: (i) poverty alleviation; (ii) improving food security and nutrition; and (iii) contributing to sustainable management of natural resources. It also makes the point that the Science Council criteria for identifying priorities considered both the effect of agricultural research on these goals and whether the CGIAR system itself should undertake the research. Like other public agricultural research systems, the role of the CGIAR system is to use agricultural research to produce public goods, although the CGIAR should generate public goods that are 'international' and relevant to multiple countries. Public goods are goods that are socially beneficial but that the private sector or other national and international research systems would not produce in sufficient quantity for lack of incentive. Considering the comparative advantage of an agricultural research system is crucial to its priority assessment.

Objectives versus measures of achieving them

Chapter 7 on priority assessment at the International Livestock Research Institute (ILRI) indicates that careful attention was paid to ILRI objectives at the start of the process. Alternative potential research themes were identified and various measures were then constructed to indicate the contribution of research themes to the objectives. This second principle, followed by ILRI, of the need to separate objectives from measures of achieving those objectives may seem obvious, but is sometimes missed by others in favour of just specifying a list of criteria and assessing research programme contributions to the criteria. The criteria may be partial measures of achieving objectives such as technology adoption rates, projected yield changes and environmental scores, but are often not applied in a way that provides consistent measures of research contributions to objectives.

Process versus tools

A third principle, emphasized in Chapter 8 for the Africa Rice Center (WARDA), is the need to separate the priority setting process from specific tools or quantitative methods applied during the assessment process. Process matters in priority assessment because: (i) priority assessment analyses do not set priorities but only provide information to decision makers who do; (ii) participation of many groups is important for buy-in to the priorities; (iii) information from multiple disciplines and from administrators as well as scientists is needed; (iv) different groups (e.g. scientists, administrators, evaluation experts) have comparative advantages in providing different types of information; and (v) some information may need to be provided, reviewed and revised in an iterative fashion to improve its accuracy and obtain buy-in.

A participatory priority assessment process is more than getting everyone involved, but requires careful assignment of roles in terms of the types of information provided or decisions made and in terms of timing of participation. Final decision makers assign weights on objectives or make trade-offs among them; scientists provide information on what is possible and technical consequences; economists design and apply economic measures of impacts, and so forth. Individual scientists can be good sources of information on specific parameters for technologies they are most familiar with, but other scientists are needed who can then check parameter estimates across multiple technologies. While scientists can be, and often are, asked what criteria they believe should be used to rank an institute's programmes and what the weights on those criteria should be, those decisions should be made considering the feasibility and distinctiveness of the criteria, and they need to embed the views of centre governance and management. Scientist views must be screened to ensure that criteria they suggest are feasible, in line with institution objectives and any weighting results in a meaningful composite measure. Process matters as much in priority assessment as the tools or methods applied.

Participatory impact pathways

The participatory impact pathway analysis (PIPA) developed by the Challenge Programme on water and food (Chapter 2) complements priority assessment in the way it elicits information in a structured, participatory manner to identify many key parameters required for priority assessment analyses and to link research to factors such as policy and other institutional constraints that might hinder impacts of research programmes. It can help make explicit the links between the research and the activities of partners and the interrelationships necessary to achieve outputs and impacts. While it is argued in the PIPA chapter that understanding of impact pathways is more important for projects than for research programmes, identifying potential impact pathways for programmes, including the interrelationships with other programmes, is also helpful in priority assessment. As the International Maize and Wheat Improvement Center (CIMMYT) chapter (Chapter 9, this volume) notes, the pathway analysis can consider the value chain from the genebank to the farmer's field and to the ultimate consumer.

Validating assumptions and learning

All of the CGIAR Center priority assessment exercises described relied on elicitation of the projected probability of research success and the adoption or influence potential of outputs delivered as a result of investment in research alternatives. In turn, these projections rely on an extensive array of assumptions about the research process, the capacity of partners in the research process, extension/implementation capacity, availability of complementary inputs, adopter preferences and market evolution. As illustrated in a number of the chapters, efforts to improve the validity of these assumptions have been an important aspect of priority assessment. As the CIP chapter (Chapter 3, this volume) illustrates, validation is aided when assumptions are followed up with evaluation and *ex post* impact assessment, which can feed into subsequent priority assessment exercises. It can be helpful if assumptions and projections are compared with historical patterns, and scientists are asked to justify a divergence between future predictions and past patterns. Their justifications can guide subsequent process evaluations. Other means of validating assumptions that are illustrated in the chapters include partner consultations (International Center for Agricultural Research for the Dry Areas (ICARDA)), the inclusion of data from on-farm reconnaissance surveys (International Crop Research Institute for the Semi-Arid Tropics (ICRISAT)) and fostering internal debate through review committees (ILRI). Geospatial analysis can be used to help define and characterize likely adoption domains, as illustrated by ILRI, the Association for Strengthening Agricultural Research in Eastern and Central Africa (ASARECA) and CIMMYT.

Reconciling multiple objectives

All CGIAR Centers and many other agricultural research systems struggle with the issue of how to establish priorities in light of the multiple objectives of poverty reduction, improvements in health and nutrition and improved sustainability of the natural resource base, and the fact that each of these objectives has multiple dimensions. Two common approaches are illustrated by the CIP and ICRISAT chapters (Chapters 3 and 6, this volume). One approach is to quantify economic benefits of research programmes and then to assess trade-offs when weights are placed on alternative objectives. The second approach is to define a large set of criteria, develop indicators or measures of programme/project contributions to the criteria and then to weight the criteria. In many cases economic criteria are included in the second (scoring) approach as well. In some cases, such as the first stage of the Center for International Forestry Research's (CIFOR) or the CGIAR Science Council's cases, criteria are provided for reference as part of an overall ranking exercise, but are not explicitly considered in the analysis.

The trade-offs approach has the advantage of providing decision makers with potentially useful information for priority assessment, while placing the onus on them to weight objectives and select the priority programmes. The scoring approach has the advantage of providing a composite ranking of programmes/projects for which it is obvious that multiple criteria were considered. The ease of participatory interaction, simplicity of presentation of the rankings and explicit listing of criteria are appreciated by participants in scoring analyses. The results of scoring, however, can be far less objective, rigorous and transparent, than are sometimes advertised. By following some basic principles described below, the trade-offs and scoring approaches can each be made more consistent, rigorous and objective without losing transparency.

Resources required

Priority assessment requires time, expertise, data and financial support, and as a result the methods employed vary by level of analysis. Priorities can be set by programmes, projects and activities within projects, but as the proportion of grant funds increases and core funds decline, it becomes increasingly important to focus quantitative priority assessment methods at the programme level. The critical decisions at this level are how much to support specific research themes, disciplines, regions and commodities with core funds to position the institute to seek grant funds for projects in appropriate areas. The expertise required to lead priority assessments at the programme level includes understanding of how to effectively elicit research programme information from scientists, training in economics and basic knowledge of biology and social sciences in general. Most of the CGIAR Centers have acquired a person with such expertise, but this person(s) must also be supported with adequate core funds to gather data, conduct interviews and interact with a wide range of stakeholders. Key issues for decision making include the selection of programmes to include in the analysis, and the size of those programmes. The unit of analysis needs to be sufficiently small that research products are cohesive and adoption/influence can be meaningfully projected.

Internationality

Several of the chapters such as the ones for the Science Council, ICRISAT, and ILRI, note that Centers attempt to provide international public goods, as part of the CGIAR mandate. Therefore, some of the Centers include 'internationality' as an explicit criterion. The concept of producing international public goods is central to the mission of each CGIAR Center because it relates to the comparative advantage of the CGIAR system. Unless the products generated are public goods, they can be left to the private sector, and unless the benefits, even if they are public goods, affect more than one country, they can be left to NARS. However, care must be exercised in how this criterion is included in a priority assessment analysis. It is not a criterion that can be applied in a linear fashion with a Center necessarily getting more credit the more the countries are affected. It is closer to being a lexicographic criterion in the sense that there must be multiple countries impacted by the public-good-type research, and if there are, the criterion is met. A Center may also want to place weight on specific regions receiving emphasis due to the unevenness of poverty around the world, but that is a separate objective from internationality.

If a Center starts weighting more countries being affected as better than a small number of countries, that measure can present a problem if it causes the Center to downgrade a programme just because it affects fewer countries, for example, if one programme affects ten countries and 1 million ha, but another programme affects seven countries and 20 million ha. Both countries clearly meet the international criterion, but a simple index based on the number of countries would downgrade the second programme in a seemingly dubious way. There is nothing significant from a welfare standpoint about a larger number of political boundaries once the international public good criterion has been met. However, there may be a political reason to prefer a larger number of countries in terms of generating support for an institution's budget.

The concept of an international public good could also relate to research issues such as how to mitigate effects of climate change, or biodiversity preservation, where the intended benefits may have international public goods characteristics, even if the adoption/influence occurs in a single country. Once again, however, the fact that the benefits are an international public good affecting more than one country is what is important.

Addressing the poverty objective

Several chapters mention, and those for four Centers explicitly include, the CGIAR objective of poverty reduction in their priority assessment analyses. The appendices to the International Institute of Tropical Agriculture (IITA) chapter (Chapter 4, this volume) and the CIP chapter (Chapter 3, this volume) indicate that they assess changes in a poverty headcount index due to research as part of their analyses. The ILRI chapter (Chapter 7, this volume) calculated poverty rates for the adoption domains of research themes but not the effects of the research on *changing* the rates. Poverty is multidimensional as described in the ICRISAT chapter (Chapter 6, this volume). As a result, the two primary methods for including it in priority assessment are to: (i) assess poverty headcount or severity indices before the research and then recalculate them after adding projected income changes to producers and consumers due to the research (see Moyo *et al.*, 2007); or (ii) assess quantitatively or qualitatively the effects on research themes on productivity of poor farmers, employment of landless labourers, wages in adopting regions, food prices and nutritional enhancement. The CIP study, for example, used a series of factors (the supply and demand elasticities, the share of output used on the farm, the share of marketed surplus consumed by poor nonfarm households, the size distribution of farms, the factor biases in the technology, the likely diffusion profiles for different socio-economic farm classes and the income distributions in affected countries) to apportion out the projected economic benefits of research programmes to the poor.

Regional and multi-organizational priorities

The International Food Policy Research Institute (IFPRI), ICRISAT, ICARDA and WARDA chapters stressed the importance of Centers setting regional in addition to global priorities. There is no reason that similar approaches cannot be used at the regional level as at the global level. The primary additional step is to prioritize across regions as well as within regions. The IFPRI and ASARECA study (Chapter 11, this volume) also considered agricultural development priorities and the need to prioritize research not just in relation to one institute but across all CGIAR Centers within a subregion. The importance of geospatial analysis of agro-ecological and socio-economic factors that influence the adoption (influence) domains for the technologies or policies is highlighted in that study. The ICARDA study made an attempt to also include NARS priorities within its own priorities, but struggled somewhat to find an adequate set of tools to assist with the participatory regional process it set in motion. The CGIAR Science Council's consultative exercise was not able to explicitly consider the impact potential of macro-level research alternatives. In general, research-institute-level priority assessment processes described in the chapters quantified more the likely benefits of specific research alternatives than did the processes that attempted to bridge larger clusters of research bodies. They also appear to have engaged internal scientists more frequently, and thereby enhanced the internal learning role of priority assessment. This suggests that there may be constraints to the effective pursuit of priority assessment at a multi-institute level.

Challenges

The CGIAR Centers and other research institutions face several challenges as they attempt to refine their priority assessment systems.

1. A key challenge in any priority assessment process is to find a means of encouraging broad participation without letting that participation undermine the integrity of the process. Participation must be structured so that people provide input to the process in line with their knowledge and roles. For example, overall goals and objectives for a public sector agricultural research institute's research programme must reflect the results of external political processes and those goals and objectives must be transmitted to the institute through its leadership team. *Measures* of achieving those goals and objectives must then be carefully crafted by someone familiar with the literature on how to construct such measures. The size of many of the technical parameters assumed when those measures are applied in a priority assessment analysis must come from those familiar with the science being considered and the context in which the scientific results will be released to and by NARS. Too often, people with different comparative advantages in a priority assessment process are brought together in a group session to discuss and rank priorities without adequate attention to *who* should be asked *which* questions. Scientists, local NARS (or other partners), institute directors, economists and others have important but different roles. Of course any formal priority assessment analysis must be supported and viewed as useful by the top management of the institution or it is a waste of resources, but it must also engage scientists in ways that promote reflection without undermining the integrity of analysis.

2. The elicitation of probabilities of research success and adoption (influence) potential from scientists whose research is being assessed creates incentives for overoptimistic projections and inadequate accounting for adoption barriers. Most *ex ante* estimates of research benefits based on these parameters have shown to be biased upwards compared with *ex post* findings (as illustrated in the CIP chapter (Chapter 3, this volume)). This means that there is need to 'validate' these assumptions through internal debate, inclusion of external stakeholders, comparison with *ex post* findings and/or other means of substantiation. If it is clear that these projections will become the basis for subsequent evaluation, some overoptimism may be mitigated. Given the importance of the learning role of priority assessment illustrated in the chapters, it is essential that validation be given explicit attention.

3. Weighting across the contributions of research programmes to a large set of criteria often appears rigorous and transparent because it is quantitative and the criteria are explicit. However, such weighting of criteria can reduce the information content of data and mislead because many criteria cannot be simply weighted and added but must be combined in specific, sometimes multiplicative, ways to make sense. Measures of contributions to individual goals can be weighted once the criteria have been logically combined, but not contributions to disaggregated criteria.

4. There are broadly accepted views on human well-being and the role of international agricultural research in contributing to that well-being. Examples of these views are: (i) human well-being is multidimensional but one goal is to improve it; (ii) more income is better than less income in improving well-being; but more so if the income reduces poverty and improves nutrition and health; (iii) research-induced improvements in productivity/efficiency favour income growth, but can have many distributional effects; some of which can make some people worse off even as it makes others better off; (iv) one distributional effect of research has a time dimension (sustainability); (v) some measures of well-being are not reflected in market prices or measured incomes, especially those affecting the natural resource base; and sustainability of the natural resource base is important; and (vi) people vary in the weights they place on distributional outcomes, but the research for development values improvements to the well-being of poor people higher than improvements to the wealthy. Any priority assessment approach for international agricultural research must address these accepted views, which can be difficult to do in a rigorous fashion without spending significant time and resources and having substantial expertise in evaluation. However, as the chapters in this book make clear, most Centers have attempted to do so.

5. Priority assessment must be cost-effective and therefore not all benefits of research can be projected in a rigorous, quantitative and transparent fashion given the current state of knowledge on impact measurement. Choices must be made about how to structure the incorporation of information on research contributions that are difficult to quantify without unwittingly misleading decision makers by presenting oversimplified results that only appear credible because they have been quantified by simple (but misleading) indicators.

6. Increased reliance on donor-restricted funds must be explicitly considered in the structure of the priority assessment process. As noted above and in the CIP chapter (Chapter 3, this volume), such reliance makes it increasingly difficult to prioritize the institution's overall programme below the level of broad programmes or the subset of projects that are core-funded, as individual donor-funded projects that fit under programme objectives can arrive at any time. However, the programme categories must be narrow enough to clearly exclude potential donor-funded research projects that fall outside the purview of the institutional priorities and to minimize overlaps in topics. Priority assessment also must occur within programmes, but those analyses tend to be more qualitative or targeted at core-funded projects for which there are questions of adding or reducing support. In some contexts, it can be unproductive to try to rank all existing and future projects within an institution from highest to lowest, in part because synergies across projects confound the analysis, and because the majority of potential projects are supported by external donors whose preferences shift over time. As noted in the CIMMYT chapter (Chapter 9, this volume), restricted funding raises the potential need to consider donors as target audiences for any priority assessment process, as follow-up may depend on their buy-in to the results. A prevalence of restricted funding can also mean that scientists who raise their own resources for research may effectively have a great deal of autonomy. This autonomy raises the importance of the 'learning' function of priority setting, as a wide pool of scientists (rather than only managers) must revise their assumptions for priorities to shift actual research practice. As a result restricted funding raises the level of scientist engagement in the priority assessment process needed.

7. Research programmes involving NRM and policy issues have grown in the CGIAR Centers and other institutes and effects of these programmes are especially difficult to quantify, especially when they may only be loosely related to specific agricultural commodities. The natural resource environment is multi-dimensional and many benefits of improving the environment are not priced in the market. Benefits of policy research, including NRM policy, are difficult to quantify because of the difficulty of establishing what would happen without the research, given the political nature of most policies.

Moving Forward

Given where the CGIAR Centers (and other research institutions) currently stand with respect to priority assessment and the challenges listed above, where do they go from here? Future priority assessment efforts for CGIAR Centers and other research institutions must strive to: (i) utilize principles, processes and methods that have been established as credible; and (ii) devote resources to improving measures of difficult-to-quantify research contributions such as those related to NRM and policy.

Utilizing established principles, processes and methods in a series of steps

A set of simple steps may help institutes make use of established principles, processes and methods in priority assessment exercises. One of the first steps is to *identify who the key decision makers are* for the institute, with respect to research priorities, and to *specify the key decisions* they need to make. Examples of basic decisions include whether to expand, contract or discontinue existing programmes, and whether to introduce new ones and if so with how many resources. In so doing, as noted in the CIFOR chapter (Chapter 5, this volume), it should also be recognized that the key decisions also depend upon implementation by a larger group of scientists who will also need to buy-in to the priority assessment results, if changes in actual research activities are to be substantive, rather than cosmetic. As highlighted by the CGIAR System Priorities chapter (Chapter 12, this volume), if there is not consensus about the full expected intent of the exercise, eventual implementation can stall.

A second step is to *define the options* to be prioritized. Early priority assessment studies focused on a set of commodity programmes in NARS, but most of the programme choices in research institutes are among non-commodity programme alternatives. The development of the 'long list' of alternatives is an important, but often neglected step in the priority assessment process. This needs to carefully blend understanding of the evolution of the external context, in terms of needs, opportunities and external capacity with understanding of current and potential internal capabilities. If the exercise is part of a longer-term strategic planning process, it can be a considerable challenge to move beyond alternatives representative of the current intentions of institute scientists to meaningfully include 'out of the box' options dependent on the evolution of institute capacity. In so doing, careful attention must be given to ensure that the alternatives to be considered are distinct, and that the unit of analysis balances the need for managing the number of alternatives compared (via broader, more aggregate units) with the unit at which impacts can most feasibly be projected (which is often at a more disaggregate 'project' level).

A third step is to *identify* precisely the *objectives* for the research institute. As noted above, for the most part these are not determined by the institute scientists but by the institution's leadership (administrators, board, and in the case of the CGIAR, 'System' leadership). Each institute will typically have 3–5 key objectives. If a scoring model is used (programmes are to be scored on their contributions to the objectives), a Delphi process can be used to elicit weights on these objectives from the institute leadership, but it is equally valid just to identify the objectives and then to provide as much information as possible about the contributions of the programmes to the objectives (and trade-offs) as information for the decision-making process.

A fourth step is to *define the measures of contributions of research programmes* to the objectives. This is the most difficult step and the one where studies often get off-track, both because contributions of some programmes are difficult to measure, and because of a desire for speed, participation and a felt need to 'quantify' contributions of every programme. This is the step where overlapping criteria can easily creep in. For some programmes, it is possible to approximate their contributions to an overall welfare or income objective or to the welfare of particular groups by designing a measure that includes information on the base value of a product, expected yield and/or cost change due to the research programme, projected rate and timing of adoption of the resulting technology or policy change, probability of research programme success and where the benefits would occur. With additional information on market parameters, economic surplus values could even be calculated for some cases. With additional information on income and consumption of representative households, a measure of changes in poverty could be constructed.

Because many measures of programme contribution are difficult to quantify, very rough measures are often included in priority assessment analyses, especially for contributions related to sustainability or the environment/health objectives. Some scoring approaches (such as those employed by ICRISAT and ILRI) appraised the predicted effects of research for criteria of environmental sustainability per unit area where adoption is predicted to occur, but did not capture the overall magnitude of environmental benefits (which depend on the area over which environmental changes will happen). One reason rough measures are used is that impacts may occur well into the future and not be priced in the market. Another reason is that 'environment' and 'health' are so multidimensional, encompassing water, air, soil, biodiversity, chronic and acute health of numerous species, and so forth. However, most research programmes primarily affect only a subset of these environmental categories. Priority assessment for those programmes can use information on the expected physical or biological effects of research programmes on specific categories. Unless non-market valuation is also conducted, so that common units (dollar values) are used, the trade-offs approach may be more useful for priority assessment involving environmental sustainability issues than is the scoring approach, as otherwise the equivalent of adding apples and oranges can occur. Many examples of successful application of scoring approaches do exist (with names such as multi-criteria evaluation framework (Hajkowitz, 2002) and analytic hierarchy process (Saaty, 2001)), and they add structure to the process, but caution is advised, as not all sets of criteria may be additive.

To keep the analysis manageable, it is best to identify only a limited set of measures of research programme contributions to the set of five or fewer centre/institute objectives. It is often more effective to devote maximum effort to developing meaningful measures and appropriate discussion of conditioning assumptions, as opposed to large numbers of criteria. This suggestion makes sense regardless of whether a scoring or the trade-offs approach is used in the eventual analysis.

The fifth step is to *obtain the data and information to parameterize (calculate) the measures of research programme contributions* defined in step four. Some of the data may come from secondary sources, but much will be obtained from the scientists themselves and in some cases from other industry experts. Small questionnaires can be constructed and interviews conducted one-on-one or in small groups. Often it is best to start with the scientist who knows the most about each programme and then to follow-up with scientists who can look across sets of programmes to ascertain which scientists may have been too optimistic or pessimistic. A group meeting may be helpful after individual interviews. Once an institute has conducted a priority assessment analysis, it has the advantage during subsequent analyses of having a database of parameters that can be used as a baseline or subjected to comparisons with what has actually happened in the field since the first analysis. Unfortunately such comparisons do not occur as often as they should in most research systems. According to the CIP chapter (Chapter 3, this volume), one reason may be the turnover in personnel who conduct evaluations, although ironically CIP is one of the few institutions that have attempted such comparisons over time with different evaluation personnel.

In step six, the *measures are calculated*. For some measures, computer programmes already exist such as the Dynamic Research Evaluation for Management (DREAM) model by Wood *et al.* (2001) or spreadsheet models. In step six, the calculated measures of research contributions can be: (i) *weighted* in a scoring model; or (ii) the opportunity cost (*trade-off*) can be *calculated* if more weight is placed on one objective as compared to another. Regardless of approach, certain objectives may not be weighted directly but can be treated in a lexicographic fashion such as the 'produces international public good (IPG) CGIAR objective. A particular research programme produces a pubic good that affects multiple countries or it does not. It is not a question of scoring (weighting) contributions to this objective as the programme must produce an IPG if it is to be supported at all, and if it does so it meets the objective. The IPG issue is one of the incentives for individual countries or the private sector to do the research if an international public institution does not, not one of size of impact (although size of impact is important in its own right).

In step seven, *results are presented* to decisions makers. If scoring is used, decision makers may want to explore the implications of changing weights. Grouping of programmes into high-, medium- and low-priority programmes can also help. If careful measures are constructed, it may also be possible to define more than one level of resource support for each programme.

Projecting Impacts of NRM and Policy-oriented Research

Results of many CGIAR and other agricultural research programmes, if successful, eventually contribute to increased production of a commodity or product that can be valued in the market. Procedures for projecting the benefits of such research programmes are well documented in the literature, both for efficiency and basic distributional effects of research (e.g. see Alston *et al.*, 1995 and many other papers). Although it is possible to measure the benefits of such programmes with varying level of detail, any research institution that does not avail itself of the established literature and models in this area is missing an opportunity to improve its priority assessment approach. However, as noted above, many types of research contributions remain difficult to measure despite previous attempts to do so. This difficulty is another reason some studies have conducted scoring exercises with only rough indicators of criteria (see Chapter 7 in Alston *et al.*, 1995, for examples). Research contributions towards NRM/environmental sustainability are one example of a type of contribution that is difficult to measure, especially *ex ante*. Contributions of research programmes that produce policy or institutional changes are another. The remainder of this chapter provides brief suggestions for projecting these two types of contributions.

Research on NRM or that which affects environmental sustainability potentially generates changes that have value through improvements in one or more environmental category. In some cases, the benefits may be realized through maintaining, over time, the value of products that can be priced in the market. Briones *et al.* (undated) provide an example for aquatic resource systems. In other cases, benefits may be realized through production of non-market goods or reductions in non-market costs, perhaps over a lengthy time period. An important step is to define the nature of the projected physical and biological effects of each research programme, and their timing. If the benefits are related to externalities that affect products that are marketed, benefits could be included in an economic surplus analysis by projecting how much costs are proportionately increased or reduced over time by the externality. The results can be discounted over time. If the benefits are non-market, it may be possible to value them with contingent valuation, experimental, cost of abatement or other methods, but the time required for applying such methods may be prohibitive unless such valuation has already been conducted prior to the priority assessment analysis. If *ex post* results from previous research are available, or become available over time, these results can inform key assumptions for future priority assessment exercises.

Benefits of NRM research can also be enumerated without valuation. An example of a physical environmental impact of research that might be enumerated without valuation would be the projected reduction in units of a toxic pesticide or the effects of that reduction on particular species or on water quality. Another would be reductions in soil loss over time. Another might be the amount of carbon sequestration. Techniques might be used to value these effects, but that may not be necessary. It also may be best not to score these effects, but simply present the projected physical or biological changes to decision makers so that they can consider trade-offs or compare them with economic benefits projected for other products of the research institute (Crissman *et al.*, 1998). If the physical effects of the projected environmental changes are valued, however, through non-market or other means, then the contributions could be scored and the implication of changing weights on objectives meaningfully explored. The common monetary units are what allow the scoring of contributions to objectives to be potentially meaningful.

For many CGIAR Centers, NRM research is one component of the larger research programme. In some cases, NRM impacts can be measured through productivity effects in a standard economic surplus model (Briones *et al.*, undated), or by projecting effects on the difference between private and social costs (Fig. 13.1) (Swinton, 2005). Research may shift either or both the private (MPC) and social (MSC) cost curves, altering the size of the external costs.

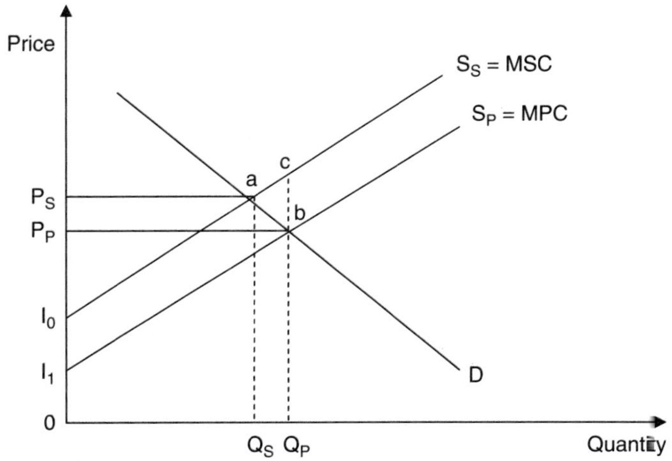

Fig. 13.1. Social supply curve lies above private supply curve when there is a negative environmental externality. The net social cost due to the externality is represented by the area abc.

In priority assessment work, it may be sufficient to ignore price elasticities (given the approximate nature of the projections), and simply assume a horizontal supply curve shifts down due to the research over time, with a vertical demand curve. The latter may also shift out gradually over time due to exogenous factors such as income and population growth (Fig. 13.2). In other words, the private supply curve may be shifting down due to the productivity effects of research, and/or the social supply curve may be shifting closer to the private supply curve due to reductions in marginal social costs associated with a change in an environmental externality. Both of these effects may need to be measured for each type of NRM impact.

There may also be NRM cases where it makes sense to simply assume a constant price (horizontal demand curve) and shift out a vertical supply curve (Swinton, 2005). These types of simple assumptions are made often in benefit/cost analyses when changes in costs are valued for a fixed quantity or changes in quantities are valued at a constant price. The more one can utilize a simple economic framework to measure impacts of different types of NRM in a priority assessment analysis, the easier it is and more credible the results become.

For institutes that are largely focused on generating NRM impacts, it may make sense to invest in quantitative models that provide detailed simulations of a variety of NRM or environmental impacts, including the effects of policies on those impacts. Effort in these models needs to be directed first at the biophysical effects of the research (even of policy research) because unless those are modelled, there may be little to say about economic impacts. One advantage of measuring NRM and/or policy research impacts in economic terms as well is that the value can more easily be used in projections of changes in poverty measures as well.

The benefits of social science or policy research associated with natural resource issues or with institutional change more generally are also notoriously

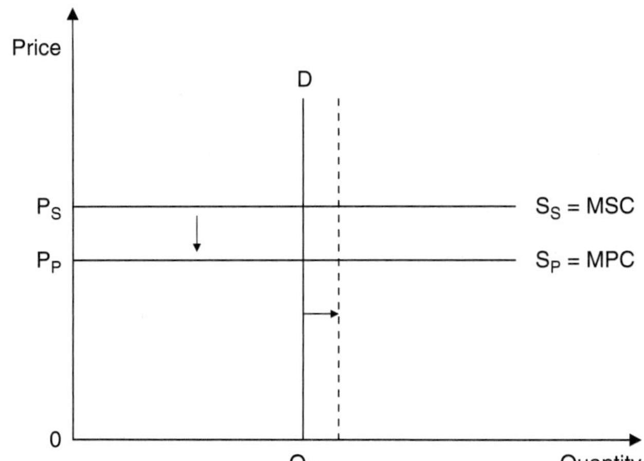

Fig. 13.2. Simplified measurement of economic impacts of NRM research.

difficult to measure even in an *ex post* sense let alone during a priority assessment exercise. Again some benefits could be captured through economic surplus measurement of the potential impacts of the policy change, and just as with improved technologies, the number and value of units of a product affected, the probability of 'adoption' of results by policy makers or firms, and timing and cost of research can be important factors to consider. An important component of the analysis is to project what would happen without the social science or policy research. A small literature has begun to develop around the topic of *ex post* assessment of social science research related to agriculture and natural resources, but none of it explicitly addresses *ex ante* analysis or priority assessment (Pardey and Smith, 2004).

The results of policy research, whether natural resource or other types, are disembodied information (Schimmelpfennig and Norton, 2003; Norton and Alwang, 2004). In *ex ante* research evaluation, projecting policy impacts resulting from research information is difficult because the policies themselves may not be evident until the research is completed, Also, if one knew the impacts of specific policies *ex ante*, there would be no need for the research. Therefore, only rough estimates of losses due to natural resource externalities and effects of NRM policy research on those losses can be used in priority assessment analysis. These estimates would have to be combined with an assessment of whether there are stakeholders who are sufficiently empowered to press for policy change. Processes of policy change can also involve many intermediate actors and parallel pathways of influence, which make *ex ante* projection even more difficult. These complications are compounded by the fact that the policy process can be discontinuous, as political change can lead to abrupt opportunities for policy reform (or reversal!) that are difficult to predict far in advance.

As noted in the CIFOR chapter (Chapter 5, this volume), there is potential to use impact pathway analysis to increase specificity about intended policy targets for research and the assumptions involved in their achievement. Network maps can illustrate the intermediate use required for policy change and the 'balance of power' between those likely to be receptive to potential recommendations and those likely to be opposed. This can provide a good basis for discussions on the probability of successful influence. A defined policy change can provide the basis for exploration of the benefits of successful influence, which in combination with defined probabilities of success can allow for projection of the expected benefits of the research. Given that precise *ex ante* analysis of the target policy change is rarely feasible within the priority assessment process, a potential 'congruence' proxy might be to assess the area affected and importance of the target policy for criteria related to research objectives, modified by the probability of successful influence.

Conclusions

This chapter has distilled and summarized some of the basic principles for priority assessment. It has argued that decisions makers are best served by a process that helps them articulate their objectives and provides them with analyses of projected contributions of research programmes to the individual objectives. In some cases, trade-offs can be presented as more weight is placed on one objective as compared to another. Caution is needed in applying scoring models that include numerous rough indicators of research contributions. Scoring can provide a means of ranking programmes, but imprecision must be recognized when weighing and aggregating criteria. Scoring is improved if measures of research contributions to the objectives are projected in common units such as dollars.

Many research programmes, especially in the natural resources and policy areas, contain impacts that are very difficult to project. In the natural resources area, maximum attention should be devoted to projecting the biological and physical effects of the technical or policy research programmes, and then projecting their value through market or non-market measures if time permits. Separating the measurement of the economic value of productivity effects from the value of externality effects of NRM research (but measuring both where relevant) is suggested as a means to projecting impacts that can be included in either scoring or trade-off analyses. It may be presently possible in some cases to project the economic value of policy changes using economic surplus analysis (see Norton and Alwang, 2004), but it is crucial to ensure that target policies and intended intermediaries are specifically identified in the process. In general, priority assessment projections are only as strong as their weakest assumptions, so it is critical that exercises triangulate predictions and foster scientist reflection, both to improve accuracy and to ensure that scientists learn through investments in information for research decision making.

References

Alston, J.M., Norton, G.W. and Pardey, P.G. (1995) *Science Under Scarcity: Principles and Practice for Agricultural Research Evaluation and Priority Setting*. Cornell University Press, Ithaca, New York.

Briones, R., Dey, M., Stobutzki, I. and Prein, M. (Undated) *Ex Ante Impact Assessment on Natural Resources Management: Methods and Application to Aquatic Resource Systems*. World Fish Center, Penang, Malaysia. Available at: http://roehlanobriones.tripod.com/papers/nrm_impact_assessment.pdf

Crissman, C.C., Antle, J.M. and Capalbo, S.M. (1998) *Economic, Environmental, and Health Tradeoffs in Agriculture: Pesticides and the Sustainability of Andean Potato Production*. Kluwer, Cambridge, Massachusetts.

Hajkowitz, S. (2002) *Regional Priority Setting in Queensland: A Multi-Criteria Evaluation Framework*. CSIRO, Australia.

Kelley, T. and Gregersen, H.M. (2005) Meeting the challenge and implications for CGIAR centers. Chapter 15. In: Shiferaw, B., Freeman, H.A. and Swinton, S.M. (eds) *Natural Resource Management in Agriculture: Methods for Assessing Economic and Environmental Impacts*. CAB International, Wallingford, UK.

Moyo, S., Norton, G.W., Alwang, J., Rhinehart, I. and Deom, C.M. (2007) Peanut research and poverty reduction: impacts of variety improvement to control peanut viruses in Uganda. *American Journal of Agricultural Economics* 89(2)(May), 448–460.

Norton, G.W. and Alwang, J. (2004) Measuring the benefits of policy-oriented social science research: evidence from two developing countries. In: Pardey, P.G. and Smith, V.H. (eds)

What's Economics Worth? Valuing Policy Research. Johns Hopkins University Press, Baltimore, Maryland, pp. 225–251.

Pardey, P.G. and Smith, V.H. (eds). (2004) *What's Economics Worth? Valuing Policy Research*. Johns Hopkins University Press, Baltimore, Maryland, pp. 225–251.

Raitzer, D. and Ryan, J. (2008) Benefit–cost meta-analysis of investment in the International Agricultural Research Centers of the CGIAR. *Agricultural Systems* 96 (March), 108–123.

Saaty, T.L. (2001) *Fundamentals of Decision Making and Priority Theory*. RWS Publications, Pittsburgh, Pennsylvania.

Schimmelpfennig, D.E. and Norton, G.W. (2003) What is the value of agricultural economics research? *American Journal of Agricultural Economics* 85(1)(February), 81–94.

Swinton, S.M. (2005) Assessing impacts of natural resource management using economic surplus. Chapter 7. In: Shiferaw, B., Freeman, H.A. and Swinton, S.M. (eds) *Natural Resource Management in Agriculture: Methods for Assessing Economic and Environmental Impacts*. CAB International, Wallingford, UK.

Wood, S., You, L. and Baitx, W. (2001) *Dynamic Research Evaluation for Management (DREAM). Version 3 Software and User manual*. International Food Policy Research Institute, Washington, DC.

Index

Page numbers in **bold** type refer to figures, tables and boxed text.